THE PRODUCT WHEEL HANDBOOK

CREATING BALANCED FLOW IN HIGH-MIX PROCESS OPERATIONS

THE PRODUCT WHEEL HANDBOOK

CREATING BALANCED FLOW IN HIGH-MIX PROCESS OPERATIONS

Peter L. King
Jennifer S. King

CRC Press
Taylor & Francis Group
Boca Raton London New York

CRC Press is an imprint of the
Taylor & Francis Group, an **informa** business
A PRODUCTIVITY PRESS BOOK

CRC Press
Taylor & Francis Group
6000 Broken Sound Parkway NW, Suite 300
Boca Raton, FL 33487-2742

Printed on acid-free paper
Version Date: 20130102

International Standard Book Number: 978-1-4665-5418-4 (Paperback)

Library of Congress Cataloging-in-Publication Data

King, Peter L.
The Product wheel handbook : creating balanced flow in high-mix process operations / Peter L. King, Jennifer S. King.
p. cm.
Includes bibliographical references and index.
ISBN 978-1-4665-5418-4
1. Production management. 2. Production control. 3. Process control. I. King, Jennifer S., 1980- II. Title.

TS155.K626 201
658.5--dc23

2012044183

Visit the Taylor & Francis Web site at
http://www.taylorandfrancis.com

and the CRC Press Web site at
http://www.crcpress.com

Contents

Acknowledgments

As the lead author of this work, I am first and foremost grateful to my daughter Jennifer for agreeing to join me in this endeavor. Her help in developing the text, in editing it and bringing a higher level of clarity, and in developing some of the mathematical background, made this task much easier and a real joy.

Neither the concepts, thoughts, ideas, and guidance presented in this handbook, nor the experience in their practical application, are solely my work. I am indebted to a number of colleagues who have helped me travel this journey:

- Bennett Foster, a valued colleague and friend, who presents frequently on the subject of product wheels. Every time I hear Bennett speak, I learn something new and useful about wheels.
- Vince Flynn, principal author of an internal DuPont™ publication, *Product Wheel Paradigms*, that influenced many DuPont plants and businesses to adopt concepts and practices similar to those described here.
- Wayne Smith, whose book *Time Out* introduced a wide variety of manufacturers to the concept of product wheels.
- Rob Pinchot, whose contributions to the DuPont Lean educational material helped me better understand product wheels, and provided a template for how to best explain product wheel design.
- DuPont colleagues Paul Veenema, Ted Brown, Lynn Mey, Kent Kimmerer, Cris Leyson, and Peter Compo, all of whom contributed to the body of knowledge and experience described here.
- John Rees's plant team at Towanda, Pennsylvania, Donna Copley's at Richmond, Virginia, Div Chopra's at Sabine River, Texas, and Greg Paris's at Louisville, Kentucky, for the valuable insights and experiences resulting from my association with them.
- Friends and colleagues at BG Products in Wichita, Kansas—James Overheul, Lisette Walker, Matt Peterson, and Gregg McCabe, who had the foresight to envision how these practices could benefit their operation, the energy and the enthusiasm to put them to work, and the foresight to challenge me when they thought we were getting too complicated.

- The people at Appleton, including Mark Richards and Steve Blasczyk, and especially Ryan Scherer, for sharing Appleton's experiences with wheels and pull for Chapter 25.
- Steve Pebly, who shared some of DuPont fluoropolymers experiences with wheels.
- Martin Fernandez, who provided insight into Dow Chemical Company's use of product wheels.
- Ray Floyd, whose excellent book, *Liquid Lean,* gave me a deeper understanding of the complexities of chemical process changeovers.
- Michael Sinocchi, executive editor at Taylor & Francis/Productivity Press, for his guidance in my novice literary effort 4 years ago, his patience when I had to slip deadlines, and his continuing support this time around.
- My daughter Courtney, herself a professional in the field of operations management and production planning, for providing insights and experiences that were invaluable to this effort.
- Finally, to my wife, Bonnie, who not only provided strong emotional support and encouragement, but also takes care of managing the affairs of Lean Dynamics so that I have time to do the fun stuff, like writing this handbook.

Pete King

Introduction

My earlier book, *Lean for the Process Industries—Dealing with Complexity*, included a chapter on product wheels, explaining the concept, cataloging the benefits, and highlighting key steps in the product wheel design process. Since its publication by Productivity Press in 2009, I've gotten numerous questions on implementation details from readers and requests for assistance from process companies wanting to design and implement product wheels. So I decided to develop this handbook to go into more detail on the design concepts discussed in my first book, and to include details on implementation and sustaining the wheels that were only very lightly touched upon the last time around.

Why Product Wheels?

The interest in product wheels stems from their ability to provide a thorough, comprehensive structure to address at least three challenges facing operations managers.

The first is the balance between the need to produce to match customer demand and the need for production leveling. A fundamental objective of Lean manufacturing is that you produce materials or products in synchronization with the rate of customer demand, known in the Lean world as takt. (*Takt* is a German word meaning "rhythm" or "drumbeat" and has been adopted by the Lean community to refer to the rate of customer demand.) If you can understand the takt for each of the products you make, and strive to synchronize the production of each product to that rhythm, you will be driving down the waste of overproduction and the cost of excess inventory. However, customer demand almost always has variability. If we have resources in machinery, operating labor, and raw materials to manufacture to the peak demand, then much of those resources will be wasted during the valleys. Lean teaches us that we should level our production to eliminate that waste. Thus we are presented with the seemingly contradictory goals of making to takt, which naturally varies, and producing at a level rate. The reconciliation is to avoid trying to match takt on an instantaneous basis, but to integrate takt over some reasonably short period of time to smooth out much of the variation so that our rate of production is much more level. A key question involves the time period over which we should integrate the variation. Product wheel design provides a very rational basis for answering that question.

The next challenge is how to schedule the full range of materials made on the same equipment or process line. In my experience almost all production lines make a variety of products, which raises questions relating to campaign strategy. Should we produce each type of product for a set period of time on some regularly repeating pattern? Or one by one, scheduled in an unstructured sequence on an as-needed basis? If we follow a regular pattern, how long should the overall cycle be, and how much of each product should we make on each cycle? Because changes must be made to the line between different products, and these changeovers are often very time-consuming and expensive, the natural tendency is to run long cycles to minimize the number of these changeovers. However, that behavior results in large inventories and high inventory carrying costs. The product wheel design process has tools to provide several perspectives on how long the overall cycle should be, what the optimum balance between inventory cost and changeover cost is, and how to account for other influences, like product shelf life and minimum practical lot sizes.

The third challenge is to minimize changeover time and material losses by optimizing the production sequence. If we're going to follow a regular pattern, does sequence matter? Is there an optimum order in which to cycle through the various products to minimize changeover difficulties? Most production planners and schedulers think they understand their products well enough to have put this question to rest, and often they're right. But sometimes they're not, and product wheel design has tools to either validate the current pattern or suggest a better one.

The ability to address all of these challenges in a thorough, comprehensive, holistic fashion is what gives product wheels their power. That has created the widespread interest in the product wheel concept and the thirst for details on how to implement them.

Process Industry Challenges

The thrust of my first book was that operations involving process manufacturing are very different in their inherent behavior from discrete parts manufacture and assembly operations, and that for Lean manufacturing efforts to be successful, they must be approached with a broader understanding of the principles. In my 20 years providing Lean guidance to DuPont™ businesses, I became more and more convinced that, to be fully effective, Lean practices must be modified to adapt to the unique challenges that process manufacturing presents. By process manufacturing I mean operations that are characterized by chemical and mechanical transformations, including chemical reactions, mixing, blending, extrusion, sheet forming, slitting, baking, and annealing. Finished products can be in solid form packaged as rolls, spools, sheets, or tubes; or in powder, pellet, or liquid form in containers ranging from bottles and buckets to tank cars and railcars. Examples include automotive and house paints, processed foods and beverages, paper goods, plastic packaging films, fibers, carpets, glass, and ceramics. The output may be sold as consumer products (food and beverages, personal care products) but more often become ingredients or components for other manufacturing processes. These operations differ enough from the manufacture of automobiles, cell phones, and refrigerators that a different perspective and a broader view must be taken for Lean and other manufacturing improvement initiatives to achieve their full benefit.

Product wheels are a dramatic example of this different perspective. Product wheels provide all the benefits that the traditional Lean production leveling technique, called heijunka, does, but also provide a much more comprehensive analysis and optimization of the scheduling challenge. Product wheels are just as applicable to discrete parts manufacturing as they are to process manufacturing, but the benefits for process operations are generally much greater. That arises from the fact that process industry product changeovers are very often more complex, more time-consuming, and more costly, so determining the optimum sequence to minimize changeover cost and the optimum cycle to balance changeover cost with inventory cost becomes much more financially important for process operations. The desire to run very long production cycles to avoid problem changeovers is seen everywhere, but has much more influence on scheduling in process operations, because they often incur expensive material losses on restart, while ramping up to specified process conditions and product properties, in addition to the more routine mechanical changeover costs.

Product Wheel Basics

The product wheel concept, conceived, pioneered, and refined within several process companies relatively independently, incorporates these features:

- Products are produced in a regularly repeating, fixed sequence.
- The sequence is optimized to minimize changeover cost, time, difficulty, or all of these factors. Products are always made in the same sequence.
- The time for one complete cycle (cycle time, wheel time) is relatively fixed.
- The overall wheel time is divided among products based on relative demand for each product. Thus higher-volume products have longer "spokes" on the wheel; lower-demand products have shorter spokes.
- Very low-demand products might not be made on every cycle, but may be scheduled for every second or every fourth cycle, for example. But when they are made, they are always made at the same point in the product sequence.
- The wheel is designed based on average product demand, but what is actually made on any cycle is based on current orders or on inventory consumed, based on Lean pull system principles.

Among the companies who have used the wheel concept to provide a competitive advantage are BG Products, Inc., Appleton, the DuPont Company, the Dow Chemical Company, and Exxon Mobil. These and other companies pioneering the product wheel methodology have used it to great advantage in the production of automotive and house paints, extruded polymers, paper and plastic sheet goods, industrial chemicals, engine oil additives, waxes and pastes, laminated circuit board materials, and a host of other products. In addition to the obvious benefits of reduced changeover cost, reduced inventories, increased capacity, and improved customer delivery performance, most users have found the greatest benefit to be the regularity and predictability it brings to the operation. They have found that the organized, disciplined structure that product wheels provide reduces the chaos often found in production scheduling, allows planners and schedulers to spend less of their personal time resolving schedule problems, and provides a stable platform so that abnormal events can be dealt with in a less stressful, more logical manner. Response to problems becomes less reactive and more purposeful.

How This Book Is Organized

The flow of this workbook is:

- Chapters 1 and 2 define the need for product wheels and the wheel concept in detail.
- Chapters 3 through 13 explain the wheel design process, and provide an example that is carried through these chapters.
- Chapters 14 through 20 define and explain the steps in implementing product wheels.
- Chapter 21 explains the prerequisites for product wheel design and implementation, improvement processes that will simplify product wheel design and make successful implementation much more likely if executed before product wheel design is begun. Appendices C through F give a more detailed explanation of some of these prerequisites.
- Chapter 22 describes how product wheels set a foundation for pull replenishment systems if designed based on the principles in this workbook.
- Chapter 23 clarifies the importance of metrics, and how certain traditional manufacturing metrics must be reevaluated so that they don't inhibit product wheel performance and limit the benefits to be derived from them.
- Chapter 24 defines the cultural foundation necessary for smooth product wheel design and implementation, and how the right approach to product wheel design can help institute and foster those cultural improvements in a synergistic way.

- Finally, Chapter 25 provides several examples of product wheels currently being used by successful manufacturing companies, such as BG Products, Inc., the DuPont™ Company, the Dow Chemical Company, and Appleton. Factors that were critical to the success of those applications are included.

Many of the specific steps in wheel design described here are not new; some are very well known and understood in the operations management community. What is new here is their application to production planning and scheduling problems, and more importantly, a process defining how and when they should be used in product wheel design. This manual provides a very detailed, step-by-step road map for product wheel development, implementation, and operation.

About the Authors

Jennifer S. King is an operations research analyst with a government contractor, analyzing operational impacts of emerging Federal Aviation Administration (FAA) technologies and developing cost and performance models to support airline investment decisions. Prior to that, she spent 5 years with the Department of Defense developing discrete event simulation models to assist the army in setting reliability requirements for new platforms, and analyzing performance of weapon systems alternatives. Her prior publishing experience includes editing textbooks and developing mathematics problems and solutions for ExploreLearning.

Jennifer has degrees in mathematics and psychology from the University of Virginia, and a master's degree in operations research from the University of Delaware. She is a member of INFORMS.

Peter L. King is the president of Lean Dynamics, LLC, a manufacturing improvement consulting firm located in Newark, Delaware. Prior to founding Lean Dynamics, Pete spent 42 years with the DuPont™ Company in a variety of control systems, manufacturing systems engineering, continuous flow manufacturing, and Lean manufacturing assignments. That included 18 years applying Lean manufacturing techniques to a wide variety of products, including sheet goods such as DuPont Tyvek®, Sontara®, and Mylar®; fibers such as nylon, Dacron®, Lycra®, and Kevlar®; automotive paints; performance lubricants; bulk chemicals; adhesives; electronic circuit board substrates; and biological materials used in human surgery. On behalf of DuPont, he consulted with key customers in the processed food and carpet industries. He retired from DuPont in 2007, leaving a position as principal consultant in the Lean Center of Competency. Recent clients have included producers of sheet goods, lubricants and fuel additives, and polyethylene and polypropylene pellets.

Pete received a bachelor's degree in electrical engineering from Virginia Tech, graduating with honors. He is Six Sigma Green Belt certified (DuPont, 2001), Lean manufacturing certified (University of Michigan, 2002), and a Certified Supply Chain Professional (APICS, 2010). He is a member of the Association for Manufacturing Excellence, APICS, and the Institute of Industrial Engineers where he served as president of IIE's Process Industry Division in 2009–2010.

Pete is the author of *Lean for the Process Industries—Dealing with Complexity* (Productivity Press, 2009), and several published articles on the application of Lean concepts to process operations. He has been an invited speaker at several professional conferences and meetings.

DuPont™ Tyvek®, Sontara®, and Kevlar® are trademarks or registered trademarks of E.I. DuPont de Nemours and Company. Mylar® is a trademark of DuPont Teijin Films; Dacron® and Lycra® are trademarks of Koch.

Chapter 1

The Problem: Production Sequencing, Campaign Sizing, Production Leveling

Product wheels have become a widely used production scheduling methodology within a few process companies manufacturing paint, salad dressings, synthetic rubbers, sheet goods, pharmaceuticals, and transmission fluids because they resolve several operational issues in a very complete, thorough, holistic way. The mixed-model scheduling techniques typically described for discrete parts assembly ignore most of the real issues found in process industry operations, so a more comprehensive approach is needed, and product wheels provide that.

Challenges Facing Operations Managers—Production Leveling

One of the key responsibilities of an operations manager is to develop a strategy for scheduling the manufacturing processes. This presents several issues, and product wheels are one very effective approach to dealing with them in an integrated, coordinated manner.

The first issue is the necessity to balance the need to produce to customer demand with the need for production leveling. On the one hand, Lean manufacturing suggests that production be synchronized with the rate or rhythm of customer demand, known as *takt*. The concept is that if you can synchronize the steps in the manufacturing process to the rhythm of customer demand, you've got a foundation in place to work toward smaller lot sizes, more continuous flow, and lower inventories, while satisfying customer demand with as little waste as possible.

But, trying to run to takt poses a challenge. Customer demand, takt, is not constant and generally has variability, as shown in Figure 1.1. If we have the resources, in labor, equipment, and raw materials, to produce to the peaks, much of that is idle in the valleys, and is thus waste. For that reason, Lean teaches us

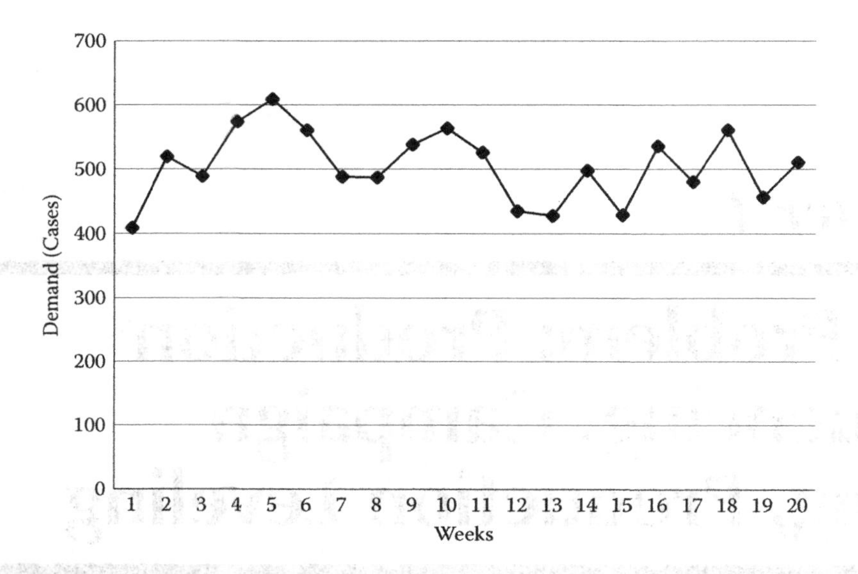

Figure 1.1 Product demand variability—One SKU.

to level our production, using techniques called *heijunka*. To quote Taichi Ohno, considered by most to be the founding father of Lean production:

> On a production line, fluctuations in product flow increase waste. This is because the equipment, workers, inventory, and other elements required for production must always be prepared for peak production.

Ohno recognized that most manufacturing operations will run much more efficiently if production is done at a uniform, level rate. This tends to maximize equipment utilization and labor utilization, and smooth out the requirement for raw materials and support facilities.

So we have the seemingly contradictory goals of (1) running at the rate of customer demand, which has natural variation, and (2) leveling out the variation in our production. The solution is to avoid trying to run at takt on an instantaneous, or hourly, or even daily basis, but to integrate the demand variability over some reasonable increment of time to create a level demand over that time increment and produce to that demand level. But what is the most practical time period over which to average out the variation? Sometimes it is obvious, but in many cases it is not. Product wheel design will help determine the most reasonable time period over which to level your production.

Challenges Facing Operations Managers—Random Sequence or Regular Pattern?

Most production operations are required to make a variety of product types on the same equipment; the large product variety required today prohibits having a dedicated line for each type. This raises the question as to whether you should cycle through the various products in an unstructured sequence, always making the most needed product (the product with the nearest ship date for which there is not enough inventory) next, or in some regular, repeated sequence.

There are several benefits in running a consistent, repeated campaign sequence, especially if the overall time to cycle through all products is also kept constant:

- A fixed pattern allows you to determine the optimum sequence, to minimize changeover difficulties, costs, and time, and then follow it routinely.
- A fixed pattern means that the same changeover combinations are seen on every cycle, so that the changeover from one specific product to another can be analyzed for improvement, with the changes practiced each time that combination comes around. The repetition of specific combinations provides a good learning platform.
- When you're producing a specific product, you know when you'll be making it again, so you know how much inventory you'll need to carry you through the period of nonproduction. You'll also know how long you will be vulnerable to variation, so you can calculate how much safety stock to carry.
- Once you've decided to run a fixed pattern, you can then decide how long the cycle should be, to optimize whatever the primary business needs are. (*Cycle*, as used here, refers to the total time required to sequence through all the products made on a piece of equipment or process line.) If lowest total cost is the most important need, the cycle can be optimized for that. If the shortest lead time is strategically important, the cycle can be designed on that parameter. Similarly, lowest inventory or highest customer performance can drive the cycle determination.
- You'll avoid the scheduling conflict that can arise if two products each have an immediate ship date that can't be satisfied with current inventory.
- Perhaps most important is the predictability and stability that this strategy brings. A fixed sequence, produced over a fixed total cycle time, allows us to know when each product will next be produced, and how much inventory is needed to carry through the cycle. Thus most of the firefighting we often see in trying to meet immediate needs goes away. The atmosphere becomes less chaotic and stressful, and more deliberate and purposeful.

Challenges Facing Operations Managers—Optimum Sequence

After you've decided to run a fixed overall cycle time, with a fixed repeating product sequence, the next challenge is to determine what that sequence should be. In many manufacturing operations, particularly those found in process companies making industrial chemicals, house paint, food products, or pulp and paper products, the time and difficulty in going from product A to product B differs from that of going from B to C, C to D, or A to D. For example, Figure 1.2 shows the products made on a sheet forming line, where the individual products can be made from different polymer raw materials, be cast into rolls of different

Product	*Designation*	*Polymer Type*	*Sheet Width (ft)*	*Basis Weight*	*Winding Tension (lb)*
A	423 J	J	12	3	250
B	403 L	L	10	3	200
C	403 J	J	10	3	200
D	423 L	L	12	3	250
E	426 J	J	12	6	300
F	406 J	J	10	6	240
G	489 J	J	8	9	220
H	406 R	R	10	6	240
I	429 L	L	12	9	320
J	489 R	R	8	9	220
K	409 R	R	10	9	270
L	409 J	J	10	9	270
M	406 L	L	10	6	240
N	426 R	R	12	6	300

Figure 1.2 Typical product lineup.

widths, have different thicknesses or basis weights, and be wound at different tensions. Changing any one of these parameters requires time, mechanical adjustments, calibration, product testing, and cost. With the variety of combinations of parameters shown, changeover from any product to any other product can be quite different in time and in cost. For example, going from product B to C requires a change in raw material polymer type, but nothing else. Going from C to D changes polymer type, sheet width, and winding tension, while a changeover from F to G changes sheet width, basis weight, and winding tension, but keeps the same polymer type. Thus it is very important to find the optimum sequence, to minimize the time and cost and allow for more frequent changeovers and improved response. Generally, the people managing the operation feel that they have this resolved, but often they do not; some kind of analytical technique to help clarify the changeover requirements and optimum patterns is usually very eye opening. Product wheel design methods provide these techniques, the specific tools needed to enable intelligent choices to be made.

Challenges Facing Operations Managers—Optimum Cycle

Having made the decision to run a regular fixed sequence, and what that sequence should be, the next issue an operations manager must face is the

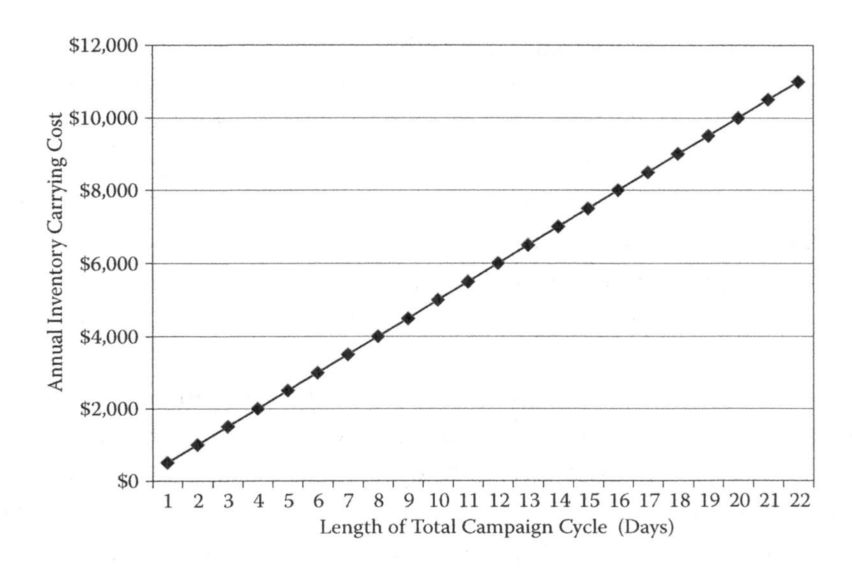

Figure 1.3 Effect of campaign length on inventory cost (one SKU).

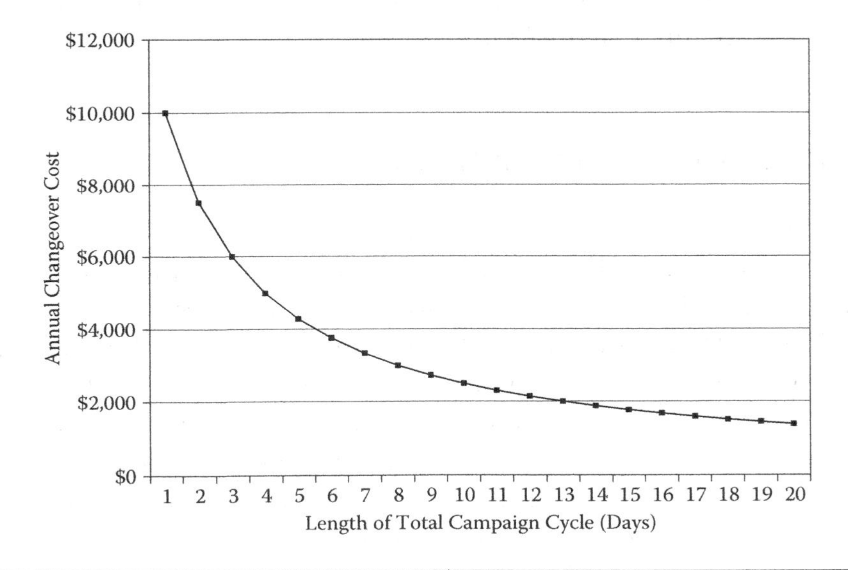

Figure 1.4 Effect of campaign length on total changeover cost (one SKU).

overall production cycle. Shorter cycles allow you to be very nimble and react to changing conditions much more rapidly. Shorter cycles need less inventory to carry you through the cycle, as shown in Figure 1.3. However, as shown in Figure 1.4, shorter cycles require more frequent changeovers, with all of the cost and problems associated with them now occurring more frequently. Appropriate production cycle sizing can balance and minimize these costs.

Therefore one of the most important steps in product wheel design is determining the optimum balance between changeover costs and inventory costs, and the wheel cycle time that will give you the lowest total cost. That cycle time may be different for each product being run on the line, and product wheel design methods allow you to reconcile those differences and arrive at a cycle time that works well for each product.

In some cases, shelf life can have an influence on production cycle time. In food processing, for example, the extensive cleanout in going from a peanut-containing product to an allergen-free product, or from a gluten-containing product to a gluten-free product, can be very time-consuming and expensive, tending to drive longer cycles. But shelf life limitations, and the retailer's requirement to hold most of that shelf life, require shorter cycles. So the cycle chosen must balance these conflicting requirements. Short-term product variability and minimum practical lot sizes are other considerations influencing cycle length, as will be seen in Chapter 8.

Product wheel design is a comprehensive way to take all of these parameters under consideration, in order to arrive at whatever production cycle time best fits the business needs. Product wheel design finds the optimum leveling period for each product, arranges the mix in the optimum sequence, and allows for the appropriate amount of each product to be made at its turn in the sequence. Thus product wheels provide an integrated, holistic, well-coordinated, step-by-step method to take all of these issues into account and arrive at the optimum scheduling strategy for any given combination of circumstances.

The Insidious Nature of Changeovers

Throughout this discussion, the term *changeover* will be used to represent any transition from one product being made to the next. It includes the time to get the line turned off and back to room temperature and ambient pressure if that is required. It includes the time to get the old material out of the system, if required. (Sometimes it is acceptable to have a minor amount of the old product left in the process, as it won't cause contamination or adversely affect product properties.) Of course, it includes the time to make any equipment changes or adjustments, and to replace any gaskets, filters, screens, and o-rings as required. What is often required, and can be the greatest time consumer and most costly step, is getting the process back to specified operating conditions, like temperature and pressure, and then getting the material being produced within required specifications after restart. In some cases this can take hours, and result in the loss of thousands of dollars of valuable ingredients.

So the changeover begins as the last molecule of the old product is being produced, and ends when the new product is flowing at the designed rate and completely within all parameter specifications. Other terms commonly used for changeovers are *setup*, *product change*, *product transition*, and sometimes the abbreviation *C/O*.

Changeovers present one of the biggest barriers to smooth flow, and one of the biggest causes of waste in time and money. If the process is running at or near full capacity, the time lost during a changeover represents revenue lost. There may be some of the previous product left in the system, which must be cleaned out, resulting in loss of that material. There are often cleaning fluids required to flush out all remnants of the old product, representing an additional cost. Then the materials lost while getting all product specifications within limits can be very costly. Below is a summary of the costs that may be involved in a changeover:

- The lost capacity and therefore revenue if the line is running at capacity
- Process ingredients from the old product lost in flushing them out
- Cleaning solvents
- Lab time and cost to ensure all contaminants are gone
- Parts consumed in the C/O: filters, screens, gaskets, o-rings, etc.
- Material lost on restart while ramping process conditions up to specified temperature, pressure, speed, torque, etc.
- Material losses after restart, while bringing product properties within specification
- Lab time and cost to ensure all product properties are within specifications

These costs and the time lost are frequently large enough to drive operations managers and production schedulers to want to make very long runs. In fact, changeover problems are often the most significant factor influencing production scheduling. The downside is that these long runs make the operation less responsive to changing conditions, make the process more vulnerable to variability, and create more inventory.

Some manufacturers have found very innovative ways to simplify changeovers. The Spangler Candy Company, producer of Dum-Dums pops, makes a Mystery Flavor™, with question marks on the wrapper where the flavor would normally be printed. This group of unspecified flavors was created as a way to continue to run the line and produce pops during flavor changeovers, with the transition material, containing a mixture of the old and new flavors, being packaged as the Mystery Flavor. So a changeover requires no shutdown, no lost capacity, and no material loss. How easy plant managers' lives would be if all manufacturers could produce "mystery" products!

Absent the ability to make mystery products, product wheels are the next best alternative. While wheels won't completely resolve the issues described above, they do provide an organized, structured way to find the optimum scheduling strategy, given the realities you must deal with. And they create an understanding of all relevant factors so that you know where to focus improvement efforts.

As you will see in the next chapter, the product wheel design process finds the optimum leveling period for each product, arranges the product mix in the optimum sequence, and allows for the appropriate amount of each product to be made at its turn in the sequence. And as you will see as you go through the wheel design process in the following chapters, the methodology is not overly prescriptive; it will not tell you exactly what to do. Rather, it will provide perspectives and guidance on possible alternatives so that you can make educated choices, while incorporating all you know about the subtle nuances of your operation.

Chapter 2

The Solution—Product Wheels

Product Wheels Defined

Now that we understand why we need product wheels, let's be clear about exactly what they are.

A product wheel is a visual metaphor for a structured, regularly repeating sequence of the production of all of the materials to be made on a specific piece of equipment, within a reaction vessel, within a process system, or on an entire production line. The overall cycle time for the wheel is fixed. The time allocated to each product (a "spoke" on the wheel) is relatively fixed, based on that product's average demand over the wheel cycle. The sequence of products is fixed, having been determined from an analysis of the path through all products that will result in the lowest total changeover time or the lowest overall changeover cost.

Product Wheel Terminology

The following terms are illustrated in Figures 2.1, 2.2, and 2.3.

- Product: Any unique material made on any step in the process. If this is the last step in the process, this will be a final product to go to a customer. If this step is early in the process, this will be an intermediate material to go to the next step. Synonyms: material, stock keeping unit (SKU), part number, D-code, item.
- Wheel time: The total time for one complete cycle through the wheel. Synonyms: production cycle, cycle time, revolution.
- Revolution: One complete cycle through the wheel, or the time it takes. Synonym: wheel time, cycle, production cycle.

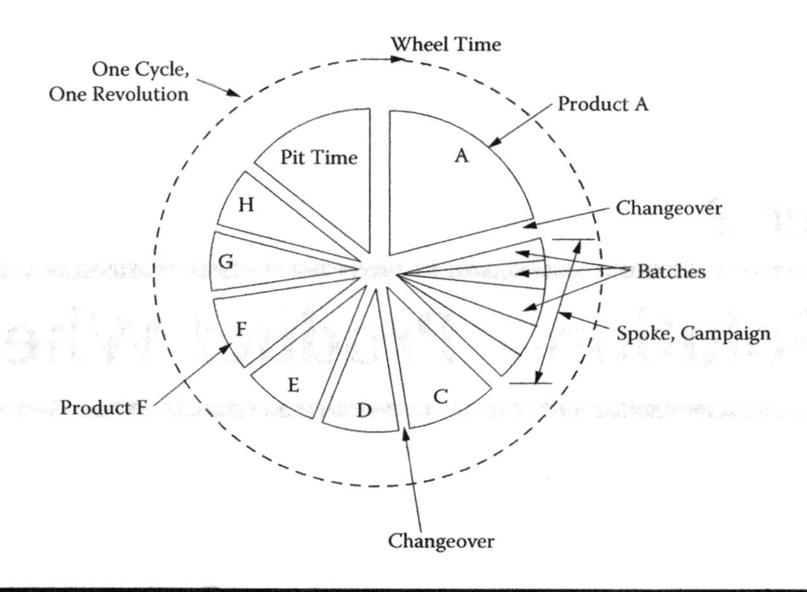

Figure 2.1 Product wheel terminology.

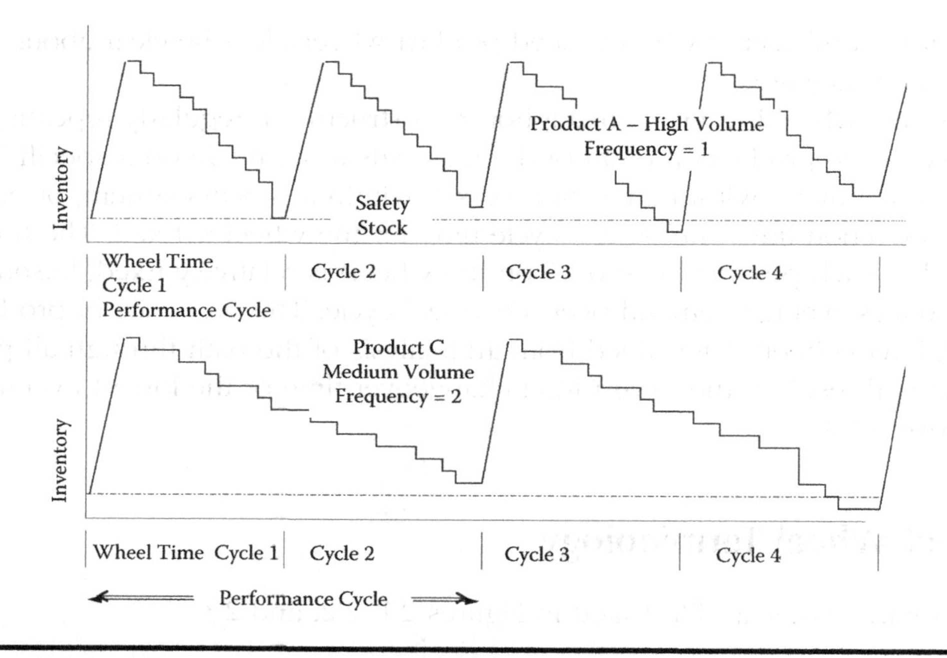

Figure 2.2 Inventory profiles—Products made on a wheel.

- Cycle: One complete run through the entire wheel. Synonym: revolution, wheel time.
- Spoke: The portion or sector of the wheel allocated to one product. It represents the time spent producing that product. If there are 12 products made on the wheel, the wheel will have 12 spokes. Different cycles of the wheel may have differing numbers of spokes; the spokes for low-volume products will be absent from some cycles. Synonym: campaign.

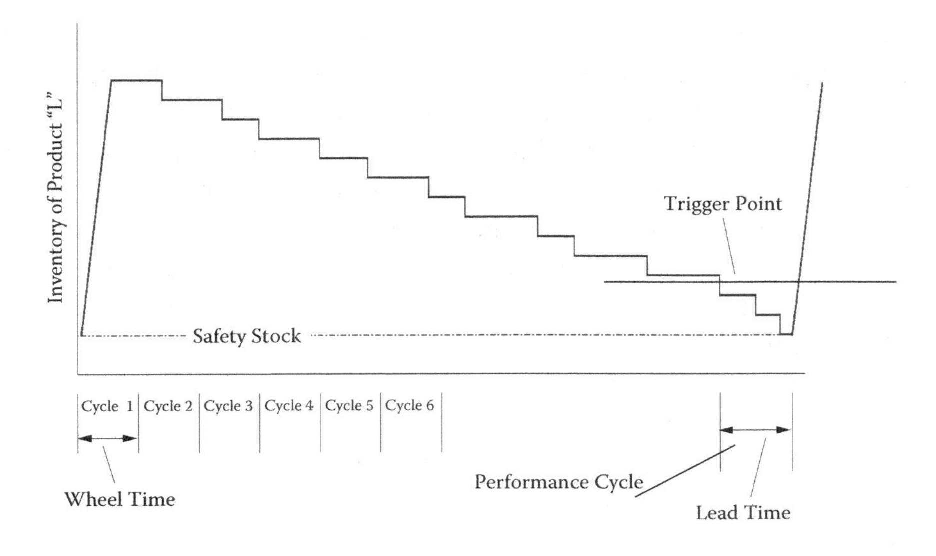

Figure 2.3 Inventory profile for a trigger point product.

- Campaign: As a noun, a batch or group of batches to be run consecutively, or the time required to produce them. The production of one product on the wheel. Synonym: spoke. As a verb, the act of grouping batches together to form a spoke.
- Batch: One lot or unit of any product being made. It is generally the smallest quantity that can practically be made. In paper production this may be one master roll; in paint manufacture it would be the quantity within one production vessel. A spoke, or campaign, may include several batches of a material. For example, the spoke for one product on a paper manufacturing product wheel may include 20 or 30 master rolls.
- Changeover: The area between spokes. The time required to terminate the production of one material, prepare the equipment for the next material, restart the process, and get to desired properties at the full design rate. Synonyms: product transition, setup.
- Process improvement time: If the required production plus changeovers doesn't completely consume the wheel time, this is the remaining time. It is so called because of its value in providing an opportunity for continuous improvement Kaizen events, operator training, new product development and qualification tests, equipment modifications, and preventive maintenance tasks. More will be said about this later. Synonym: PIT time.
- Frequency: For the lower-volume products not made on every wheel cycle, the number of cycles over which this product is made one time. For example, the frequency would be three for a product made on every third wheel cycle.
- Trigger point: The inventory level for a specific product that triggers that product to be made on the next wheel cycle. There may be some low-volume products that are made to stock (MTS), not made to order (MTO), but are low enough in volume that they are not made on any set frequency, but

only when the inventory gets below a certain limit, called the trigger point. The trigger point is calculated from the expected demand during the lead time plus some safety stock needed when demand is higher than expected.

- Performance cycle: A term used in inventory calculations that represents the time a product is at risk of stock-out due to variabilities. With product wheels, it is the wheel time multiplied by the frequency. As an example, on a 1-week wheel, a product made every second wheel would have a performance cycle of 2 weeks; that product is vulnerable to stock-outs for 2-week periods at a time.
- FSVV production: Fixed-sequence variable-volume production—another name for product wheels.
- Rhythm wheel: Another name for product wheels.
- Color wheel: Another name for product wheels, often used in paint manufacture.
- Wheel breaker: A significant event that will cause the person in control of the schedule to want to deviate from the wheel design. A typical example is a potential stock-out due to unexpected abnormally high demand for a product. Anticipating wheel breakers and creating contingency plans is a very important step in wheel design and is discussed in Chapter 17.

The spokes of the product wheel can have different lengths, reflecting the different average demands of the various products. High-demand products will have longer spokes than lower-demand products.

The amount of any product made on a specific revolution may be slightly more or less than the average demand on which that spoke was designed; because the demand on any cycle can be slightly more or less than the average demand, the amount actually produced will equal the actual demand for that material since the last cycle. Thus the width of each spoke can vary from cycle to cycle based on actual demand, but the total wheel cycle time will remain fixed.

A number of factors can influence the determination of overall wheel time. If having the shortest possible lead time through the operation is paramount, then the shortest wheel time that allows for all required production and all necessary changeovers may be selected. If the lowest manufacturing cost is a key driver, then the wheel time can be calculated to give the best balance between changeover costs, which decrease with longer wheels, and inventory cost, which increases with longer wheels. Other factors, such as product shelf life, short-term variability, and minimum lot size requirements, may also come into play, as explained in more detail in Chapter 8. The most important requirement is that once the wheel time has been selected, it must remain constant until conditions change enough to require a recalculation of wheel parameters. (The need to rebalance the wheel periodically is covered in Chapter 20.)

High-volume products will have the major influence on wheel design. The wheel time will generally be selected based on what makes most sense for these high-volume products. Lower-volume products will be made at a frequency that makes sense based on their volume; they may be made on every second, or every third, or every fourth cycle. Very low-demand products, particularly those with highly variable demand, might not be made on any regular frequency, but only made when specific orders come in. Most products on the wheel will be made to stock (MTS), but these low-volume, high-variability products are best made to order (MTO). One nice feature of product wheels is that MTO products can coexist with MTS products quite nicely. Even though the low-volume and the MTO products are not made on every wheel cycle, it is still important to determine where they fit within the overall sequence, and when they are made, always make them at that point on the wheel cycle.

Product wheels support a pull replenishment model. That is, the wheel will be designed based on average historical demand or on forecast demand for each product, but what is actually produced on any spoke is just enough to replenish what has been consumed from the downstream inventory, in accordance with Lean pull principles.

Simultaneous Operating Modes

A single wheel may simultaneously have products made under four different strategies:

- Made on every wheel cycle, so the frequency is 1. This will be the case for each of the high-volume products
- Made on alternate cycles, with a frequency of 2, 3, or 4. This will be the case for products with medium volume, with enough demand to be made frequently, but not on every cycle. A key benefit of product wheels is that each product is made at a frequency that works best for that product.
- Rather than being made on any specific frequency, made only when the inventory for that product drops below a trigger point. This may be the case for very low-volume products that can't be MTO for some reason, such as customer lead time commitments.
- Made only when there is a specific order—MTO. This is the best strategy for very low-volume products, especially those with unpredictable demand. MTO requires that the lead time commitment to the customer for those products be long enough to allow the product to be made on the wheel at the appropriate point in the sequence. The process for determining which products are good candidates for MTO is described in Chapter 6.

Product Wheel Characteristics

To summarize the key characteristics of product wheels:

- A regularly repeating sequence of the production of the various products or materials made on a specific piece of process equipment or on an entire production line.
- The sequence is fixed.
- The sequence is optimized for minimum changeover loss: time, material, and/or cost.
- The overall cycle time is fixed.
- The cycle time is optimized based on business priorities—typically shortest lead time or minimum total cost, but other factors such as shelf life limitations are also considered.
- Spokes for the various products will have different lengths, based on the average demand for each product.
- The length of any spoke, i.e., the amount actually produced on that spoke, can vary from cycle to cycle, based on actual demand for that product during the previous cycle.
- Some low-volume products may not be made every cycle.
- When they are made, it's always at the same point in the sequence,
- Make-to-order and make-to-stock products can coexist on a wheel.
- Product wheels support a pull replenishment system.
  - Each spoke is designed based on average historical demand.
  - But because actual demand on any cycle can be greater or less than average, what is actually produced on any spoke is just enough to replenish what has been consumed from inventory over the past cycle.

Process Improvement Time

There often is more time within a wheel revolution than is needed for production and changeovers. A common term for this extra time is slack time, which is a very inappropriate term. Referring to this as slack tends to de-value the time and cause people to treat it casually. This is time that can be extremely valuable to the operation, for:

- Preventive maintenance tasks
- Installation of a new piece of equipment, or major modification to an existing component
- New product development trials
- New product qualification runs
- Operator training
- Kaizen events to conduct 5S practices

- Kaizen events to reduce changeover time and cost, using single-minute exchange of dies (SMED) or other techniques (SMED is described in Appendix D)

Referring to it as process improvement time rather than slack time tends to highlight how valuable this time is, and creates a mindset that it shouldn't be wasted.

Process improvement time is usually shortened to *PIT time*. I realize that the term *process improvement time time* is redundant, but teams seem to think that *PIT time* is a more appropriate description than *PI time*.

Benefits of Product Wheels

Product wheels offer a number of benefits when compared with other scheduling strategies. Most importantly, they satisfy all of the needs discussed in Chapter 1. Specifically, they:

- Level production within the shortest practical time periods.
- Provide tools to determine the optimum sequence of products.
- Provide several relevant perspectives to allow overall production cycle time to be set on a knowledgeable, informed basis.
- Find the best economic balance between total of inventory cost and changeover cost.
- Give a firm basis for determining inventory requirements. Because we know when we will be making a given product next, we can determine how much inventory will be needed to support demand until the next cycle. We also know how long we will be at risk to variation, and thus have a logical basis to calculate safety stock.
- Provide a defined, predictable time (PIT time) for preventive maintenance tasks, equipment additions and modifications, operator training, 5S and SMED activities, and new product qualification runs.
- Provide a more stable, more structured foundation for the production schedule so that unexpected events can be accommodated with less chaos and stress.

All of this generally leads to increased usable capacity, lower inventories, and improved customer delivery performance.

Product Wheel Applicability

Product wheels may be applied to an entire production line, such as a salad dressing bottling line, a synthetic fiber spinning operation, or a frozen pizza manufacturing/packaging line. Or, they may be applied to a single large piece of process equipment, such as a plastic pellet extruder, a nonwoven fabric forming machine, or a resin reactor used in paint manufacture.

Product wheels should be considered for any piece of equipment, vessel, or process system where changeovers have costs associated with them, and especially where these costs are dependent on the specific sequence followed. Chapter 5 provides additional guidance on deciding where to apply product wheels.

Product wheels are best implemented as part of an integrated, well-thought-out Lean plan, coupled with standard work, visual management, value stream mapping, total productive maintenance (TPM), changeover improvement, bottleneck optimization, and virtual cellular flow patterns. The need for this integration, and the synergies it provides, are covered in more detail in Chapter 21.

Chapter 3

The Product Wheel Design and Implementation Process

The best way to more thoroughly comprehend the concept of product wheels and appreciate the power they provide is to understand the steps in product wheel design. Experience has shown that following this road map, and fully executing each step, will result in the most effective, most sustainable wheels.

Product Wheel Design

Step 1: Begin with an up-to-date, reasonably accurate value stream map (VSM).

Design of product wheels requires two general classes of information, one describing the throughput and performance of the equipment, and the other describing the product lineup, including demand history and trends and product characteristics affecting changeovers.

A VSM, described in detail in the next chapter, gives the former. It provides a holistic view of how material flows through the process and how it is transformed from raw materials into the end product. It includes data that indicate how smoothly material is flowing, such as reliability, yield, throughput, demand (takt), changeover times and losses, and how many products are processed at each step. It shows in-process inventories, their quantities, and days of supply. Thus it provides most of the data required to design product wheels.

More importantly, it provides perspectives and specific data that are very useful in deciding where in the process product wheels would be appropriate.

Step 2: Decide where to use wheels to schedule production.

Product wheels can be used to schedule an entire production line as an integrated unit, which is often done with packaging lines. They can also be applied to a single piece of process equipment, or to two or more pieces of equipment in series in a line. So a decision must be made on where wheels are appropriate and beneficial, and the VSM provides very useful information for that.

Once we have decided where to apply wheels, the remaining steps must be done for each specific wheel based on its requirements and operating parameters.

Step 3: Analyze product demand volume and variability—identify candidates for make to order.

The next step is to analyze the volume of demand for each product processed on this line or piece of equipment, along with the variability in demand. This will be used to determine which products should be made to stock and held in inventory, and which should be made only in response to a specific order.

Step 4: Determine the optimum sequence.

As we saw in Chapter 1, changing from one product to another can take different times and have different losses, depending on the product you're changing from and the product you're changing to. The wide variety of products produced on any process line can have several parameters that must change leading to different changeover times and losses, depending on which parameters must be changed and how much they must be changed. Thus one key to minimizing changeover losses and designing the most effective wheels is to make sure you have the optimum sequence through all the products that must be made.

Step 5: Analyze the factors influencing overall wheel time.

This is one of the most important steps in the whole design process, and a number of factors must be taken into account.

1. The first factor to be considered is the fastest wheel that can practically be run. Running the shortest, fastest wheel offers a number of advantages. It may not be the most economical choice, and it may not be the best choice, but it is useful to know what the shortest wheel time could be.
2. Next we want to know what the most economical wheel time would be, balancing changeover costs with inventory costs. Again, this is not always the final answer, but this factor generally has the most influence on wheel time.
3. If product shelf life has limits, this must be taken into account. With some food products, for example, this can force shorter wheels.

4. In some cases, product demand variability is very large when taken in very short increments, but smooths out over longer time periods. This sometimes requires longer wheel cycles than we would otherwise want.
5. In some situations, there is a minimum lot size that can practically be made. This can sometimes stretch out the wheel time; more often it decreases the frequency of the lower-volume products but not the overall wheel time.

Understanding all of these factors is vital to arriving at an optimum wheel time tailored to your specific process and business needs.

Step 6: Determine overall wheel time and wheel frequency for each product.

From the perspectives developed in executing step 5, we must now decide what the overall wheel time will be. We could choose the fastest wheel, or the most economical wheel, or some intermediate compromise. We must also consider the other factors discussed. This is not a straightforward decision, as there are advantages and disadvantages with each possible choice. But there should be enough useful information from step 5 to make an intelligent choice.

Having selected the primary wheel time, we must then set the frequency for each of the lower-volume products. This is generally straightforward, and is most influenced by factors 2, 4, and 5 above.

Step 7: Distribute products across the wheel cycles—balance the wheel.

Now that the wheel time has been determined, along with the frequency for all the lower-volume products, the next step is to decide on which specific cycle each product with a frequency greater than one should be made. For example, a product with a frequency of 3 could be slotted in on cycles 1, 4, 7, and so on, or on cycles 2, 5, 8, etc., or cycles 3, 6, 9, etc. These lower-frequency products should be slotted in a staggered fashion so that the total planned production is relatively balanced from cycle to cycle.

Step 8: Plot the wheel cycles.

Now that the specific makeup of each wheel cycle is set, there is significant value in preparing diagrams, in pie chart or bar chart format, to illustrate the various cycles. This visual representation gives all involved a much better understanding of the concepts, the design, and the relative values. It is far easier to get a feel for how much of the wheel is allocated to each product from a pie chart than from a table of demand and volumes. These charts are particularly valuable for people who weren't involved in the design details, by giving them a much better sense of how the wheel will actually operate.

Step 9: Calculate inventory requirements.

Managers will always want to know how much inventory is required to support the wheel, so that they can agree that the correct balance has been struck between inventory and changeover frequency, or suggest modifications. The latter will cause looping back to step 5 or 6. To reduce the likelihood that modifications will be needed, rules of thumb can be employed during step 5 to get an approximate feel for total inventory required, so that any concerns can be addressed then. So ideally this step is just a more exact confirmation of those approximations.

Step 10: Review with stakeholders.

It is critical that everyone who is affected by product scheduling understands how this will change his or her part of the operation. This includes people involved in actual production as well as maintenance, test lab operations, accounting, warehousing, marketing, and others. Clearly, the more that these people are kept on board during the design, the less we will need to repeat prior steps.

The visual representations developed in step 8 are very helpful here.

Step 11: Determine who "owns" (allocates) the PIT time.

The concept of process improvement time (PIT time) was touched upon earlier. This is time that is not required for production or changeovers; it is available for other valuable activities such as preventive maintenance, operator training, new product qualification runs, and continuous improvement events aimed at equipment performance improvement. To ensure that it is used most effectively, a specific person or team should be given responsibility for reconciling all needs for this time and allocating it appropriately.

Step 12: Revise the scheduling process.

None of this will work very well unless the wheel design, schedule, and all operating parameters are incorporated into the plant enterprise resource planning (ERP) system, so this is a critical step.

These steps complete the product wheel design. Once they're done, it's time to focus on implementation.

Product Wheel Implementation

Step 13: Develop an implementation plan.

A good design won't actually make anything happen. There are a number of tasks required for successful implementation, and they must be planned, scheduled, and staffed.

Step 14: Develop a contingency plan.

There are generally a number of unexpected events that cause people to want to break out of the wheel, i.e., to deviate from the planned sequence or spoke timing. While these situations can't all be fully anticipated, some can. It is very worthwhile to engage planners, schedulers, and manufacturing supervisors and operators in a discussion of what could go wrong, how to recognize it early, what steps to take to remedy it, and if the wheel must be broken, how to get back on the wheel quickly and smoothly.

It's far more purposeful to do this planning in advance rather than in the middle of a crisis. And thinking your way through all of the events that you are able to anticipate develops some principles and sets a pattern for dealing with the events that you can't anticipate.

Step 15: Get all inventories in balance.

It is rare that the inventory for each product is at the correct level when it's time to turn the wheel on; generally, inventory for some products is high, while others may be low. High inventories won't be a problem; the wheel operation will naturally bring them down, by not running those spokes until they reach the targeted level. However, the low inventories must be brought up to the target before the wheel starts or stock-outs will occur.

Step 16: Put an auditing plan in place.

There may be subtle factors that weren't taken into account in the initial wheel design, so performance must be monitored to ensure that the wheel is operating as intended. A plan should be in place to collect and analyze inventory levels, stock-out frequency, customer service performance, wheel-breaking events, and other key parameters on a routine basis.

Step 17: Put a plan in place to rebalance the wheel periodically.

Things rarely stay constant in any manufacturing environment: demand for some products may increase while others may fail, and equipment performance may improve. Wheels are quite tolerant to minor changes, but must be rebalanced when major shifts occur. Rather than waiting until these shifts have adversely affected wheel performance to recognize them, there should be a plan in place to reexamine all wheel premises periodically to anticipate a need to rebalance. Being proactive rather than reactive will allow for much smoother transitions to new process or market conditions.

These are all the tasks required for product wheel design and implementation. The next chapters give a much more detailed description of each step, using the process described in the next chapter as an example. Following these steps in the recommended sequence will get you ready for wheel turn-on in the shortest reasonable time and result in the most effective wheels.

Product wheel design is not an exact science. At several steps in the process the methods used will not give you a perfect answer, but will provide useful perspectives to enable an informed, educated decision to be made. For example, if you wish to base the overall wheel time on the best balance between changeover costs and inventory costs, the calculations used will generally give a different optimum wheel time for each product. But the numbers produced should allow you to make a compromise that is very reasonable for each specific product.

Although the process described above is presented as being very linear, wheel design is actually an iterative process. What you learn in one step may provide new insights and alter your thinking about a prior step. And the decisions you make in a given step may alter some of the mathematical assumptions made in an earlier step. For example, in performing the calculations in step 5.2, you may decide that several low-volume products will be made every fourth cycle, rather than every third. Knowing that, you can go back to step 5.1 and recalculate the shortest possible wheel time, which will now be shorter, with some products removed from each cycle. Or if you've planned for a relatively short wheel, you may learn that the QC lab can't handle the increased volume of testing required by the more frequent changeovers, forcing a loop-back to steps 5 and 6 to increase the wheel time. The amount of looping back will generally be reduced by including all relevant perspectives in the design from the start. Kaizen events are an excellent way to make this happen.

Kaizen Events

Product wheel design and implementation is best done as a team process, so that the viewpoints of all affected groups can be represented. Including more input in the initial design reduces the amount of iterative looping back that must be done once an initial design has been put on the table. If your plant uses kaizen events as a way to get improvement done quickly, design of a product wheel makes an excellent kaizen scope.

The word *kaizen* is typically used to represent two different, but very aligned processes. It is used to signify a culture of continuous improvement, where everyone in the operation has the responsibility to find ways to make incremental improvements to the work they do on an ongoing, continuous basis. The idea is that thousands of minor incremental improvements add up to very large improvements over time. Kaizen is also used to describe an event where a small team is given a process step to improve, and is then totally dedicated to that task for a few days. This is not an academic exercise; in addition to data gathering, analysis, and redesign of the work process, implementation is actually done during the event or very shortly thereafter.

The typical kaizen event is 3 to 5 days in length, but can be much shorter; depending on the scope, a kaizen event can be successfully completed in a day or less. Kaizen events to design product wheels are typically in the 4- to 5-day range.

Thus kaizen events are very focused, team-based activities that provide a very powerful way to make improvements quickly and in a way that engages all stakeholders. The benefits include:

- Things get done. Improvements are accomplished. Waste is eliminated.
- Results are seen very quickly.
- Participants learn Lean tools and get experience in their application.
- Participants gain a better understanding of other parts of the manufacturing operation, due to the cross-functional nature of kaizen teams.
- The solution is owned by the creators, and so has greater likelihood of being sustained.
- A successful event builds energy and momentum for the entire Lean effort.
- Employee motivation and morale is improved.

A kaizen event is therefore a very effective way to design and plan implementation of a product wheel. It will increase the likelihood that all relevant points of view are represented as the design proceeds, will generally reduce the need for looping back, will likely result in a better design, and get it done rapidly and deliberately.

Prerequisites for a Product Wheel

There are a number of things that should be done before beginning any serious product wheel thinking. They are highlighted in Chapter 21 and described in more detail in Appendices C, D, E, and F. Doing these things prior to wheel design forms a solid, stable foundation on which wheels can be designed and implemented, and will generally result in simpler, more effective wheels.

Chapter 4

Step 1: Begin with an Up-to-Date, Reasonably Accurate VSM

Product wheel design should always start with a thorough, complete flowchart of the entire manufacturing process. Although this doesn't have to be in the form of a value stream map, a VSM does satisfy all of the needs very well.

1. A view of the entire process, with key flow data on each significant piece of equipment, is needed to decide where product wheels might be appropriate.
2. The data boxes on the VSM provide the detailed information needed to decide if any given step is a candidate for a product wheel.
3. For any step that will be scheduled by product wheels, the VSM data boxes give the data needed for specific steps in wheel design.
4. The information flow gives an overview of the current scheduling process, which highlights which information processing steps might be modified to put the product wheel plan into practice.

An Example Process—Sheet Goods Manufacturing

To demonstrate the principles being explained and illustrated in this chapter, and throughout the rest of the handbook, a multistep process (Figure 4.1) that manufactures rolls of paper-like sheet goods will be used.

The process begins with raw materials in the form of polymers, viscous plastics that are melted and extruded as very fine fibers. The fibers are extruded onto a wide belt, thousands at a time, to form a molten web of material perhaps 10 to 12 feet wide. As the molten polymer fibers cool, they form a solid sheet, which is wound up to form a roll 10 to 12 feet wide and perhaps 4 to 5 feet in diameter. This step is called sheet forming and produces master rolls.

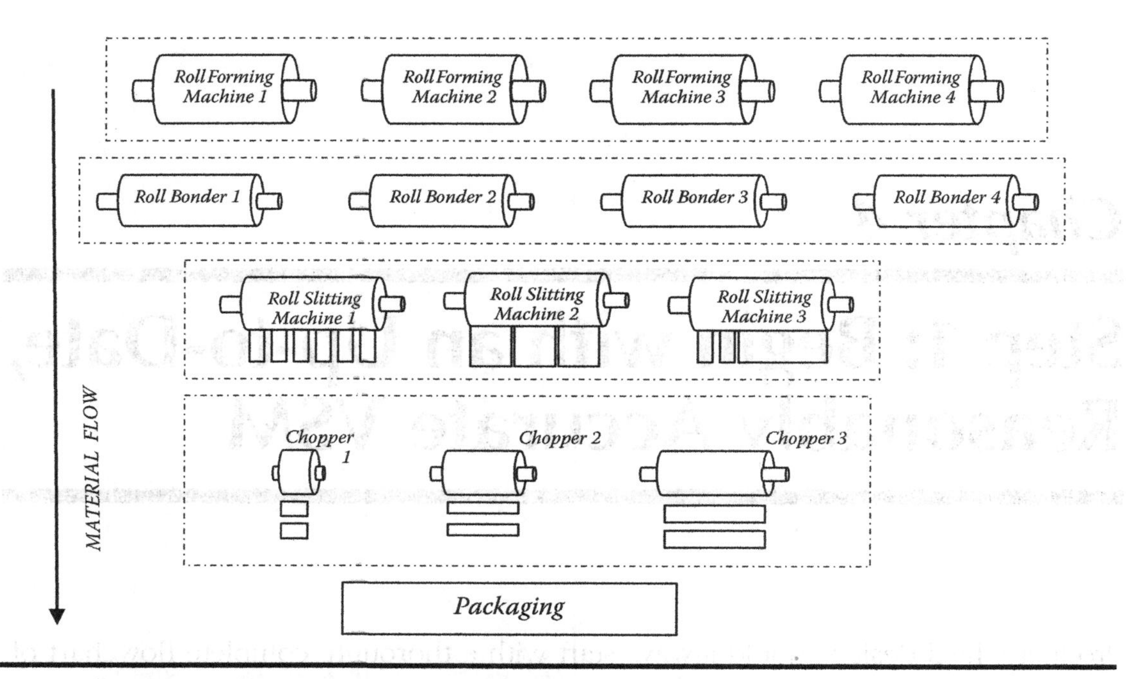

Figure 4.1 Block diagram—Sheet goods process.

The next step is called bonding, which is a form of heat treating to heat-set or solidify the sheet properties. A master roll is unwound and processed over a large heated roll; guide rolls are configured so that the sheet will be in contact with the heated roll over approximately 75% of its diameter. The bonded rolls are then wound up into rolls similar in size to the master rolls.

The next step is slitting, which cuts the wide roll into narrower rolls to meet end use requirements. The bonded roll is unwound and passed over a roll with several rotating knife blades positioned across it, to cut the 12-foot-wide roll into six 2-foot-wide rolls, for example. Bonded rolls are typically slit into 1-, 2-, 3-, 4-, and 6-foot widths to match individual customer specifications. Some rolls are sold in their original cast widths, and slitting is used simply to trim the rough edges off. The slit rolls are then wound up to go to chopping.

Chopping is the name of the process step that cuts the sheet material in the transverse direction, across the width. Slit rolls are unwound and cut across the width to form either shorter rolls, which will be wound up as chopped rolls, or individual sheets, again depending on end use requirements.

The sheets or chopped rolls are then packaged, boxed, palletized, and sent to a warehouse for storage awaiting customer orders. This is a complete make-to-stock operation, and enough material is expected to be in storage to satisfy any order and to be shipped within 4 days.

Thus there are five major steps in the process: sheet forming, bonding, slitting, chopping, and packing. There are very significant inventories in between each process step, and master rolls, bonded rolls, and slit rolls are typically stored in an automatic storage and retrieval system (AS/RS), also known as a high-rise stacker crane system. Chopped rolls and sheets are wrapped and stored in finished product inventory.

Several pieces of equipment are needed at each step to meet the high production volumes required. Prior to the introduction of Lean production methods, material flow patterns tended to be purely random, with master rolls from one forming machine going to any of the bonders, and bonded rolls from a specific bonder being sent to any slitter. As we will see in the next chapter, virtual cells, with dedicated flow lanes, will be applied to this process before any product wheel design will begin.

While the specific process being described makes a generic paper product, very similar processes are used to make Tyvek® house wrap, paper for computer printing needs, x-ray films, plastic food wrapping films, web material for hospital gowns and curtains, and plastic films for industrial uses. The specific operations being performed may be different, especially in step 2 (bonding), but the equipment footprint and flow patterns are very similar. And the performance of the equipment and the wide variety of products are extremely similar in ways that affect product wheel application.

A Value Stream Map

Figure 4.2 is a complete value stream map (VSM) of the sheet manufacturing process just described.

Value stream mapping is based on Toyota's material and information flow diagrams, and provides a very effective framework for depicting the process in a way that highlights waste and the negative effect it has on overall process performance and flow. As popularized by Rother and Shook (2003) in *Learning to See*, it has become a standard way to describe flow, and the starting point for many Lean initiatives, especially those including product wheels.

A VSM consists of three main components:

1. Material flow: Shows the flow of material as it progresses from raw materials, through each major process step (machine, tank, or arrangement of vessels), to finished goods moving toward the customer. This is a high-level view showing only major pieces of equipment or processing systems. All inventories along the flow are also shown.
2. Information flow: The flow of all major types of information that govern what is to be made and when it is to be made. This starts with orders from the customer, moves back through all significant forecasting, planning, and scheduling processes, and ends with schedules and control signals to the production floor.
3. Time line: Shows the value-added time and contrasts it with non-value-added time. It appears at the bottom of the VSM in the form of a square wave. This is a key indicator of waste in the process. It shows the effect of waste but not the cause; that should be diagnosed from the other two components of the VSM.

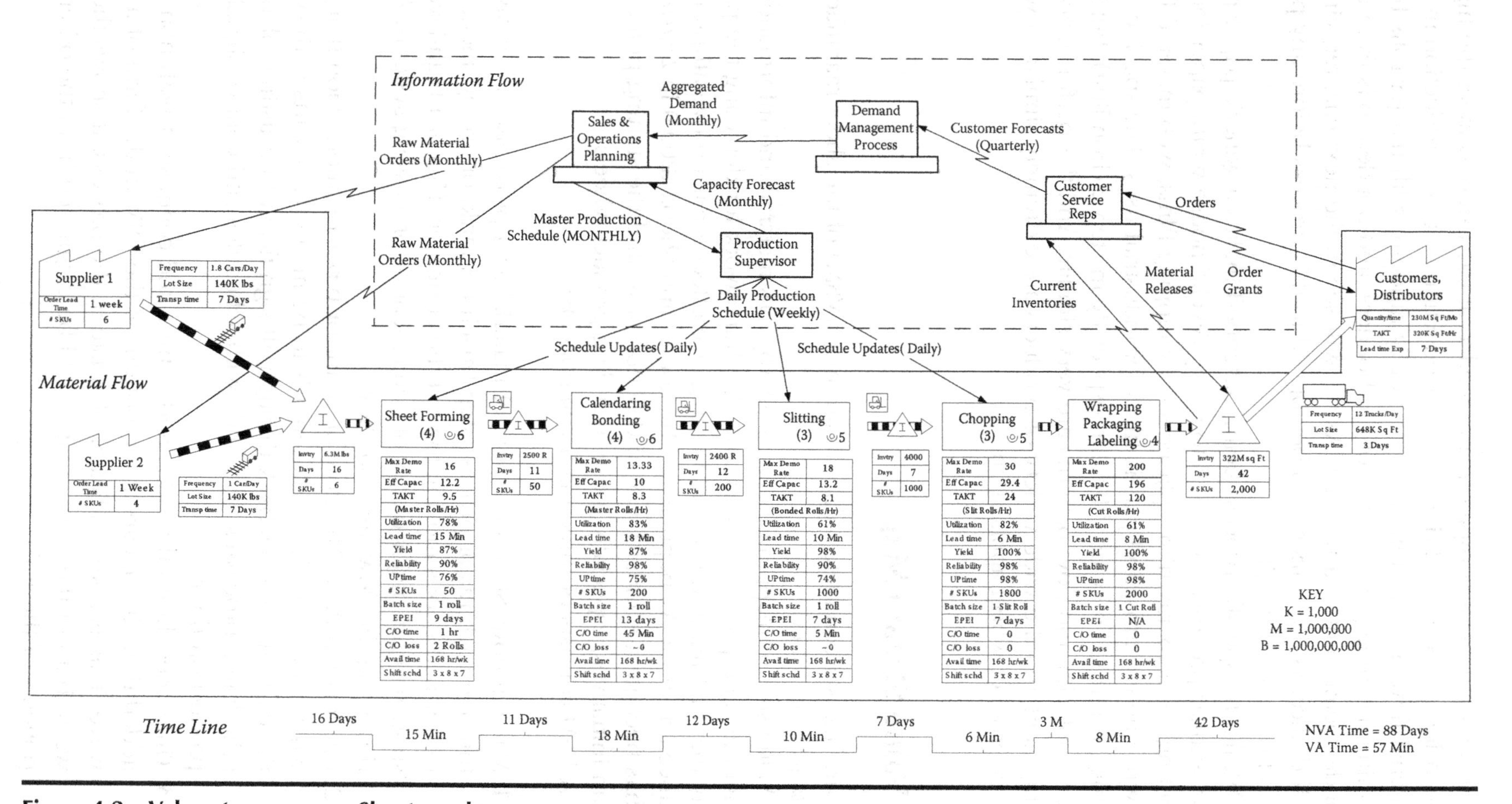

Figure 4.2 Value stream map—Sheet goods process.

The material flow and the information flow are the components that are most useful in product wheel design.

Material Flow—Process Boxes

Each major step in the process will be described in the material flow component by a process box. A process box may depict a single large machine or chemical process, like a bonding machine, a carpet tufting machine, or a paint mixing vessel. It could also depict a step in the process consisting of an integrated process system. A carpet dyeing system consisting of dye mix tanks, heaters, pumps, and the dye application machine could be shown as a single process box, as could a continuous chemical polymerization system with several tanks and much process piping.

If there are several identical or nearly identical pieces of equipment in parallel at any step, they can be shown by one process box, with the number of like units shown in parentheses. Our sheet VSM, for example, shows that we have four forming machines, four bonders, three slitters, and so forth.

Process Step Data Boxes

Each process box is accompanied by a data box, which provides the numerical information required to understand how well material is flowing through the process, where bottlenecks or capacity constraints exist, and where waste exists in the process. This information may provide clues to the root causes of delays, bottlenecks, and waste. Data boxes quantify how well material is flowing, and highlight barriers to smooth flow.

A process box and data box for sheet forming machine 1 is shown as an example in Figure 4.3. Most of the terms used in the data box are self-explanatory, but some may not be:

- Takt: As defined in Chapter 1, the rate of demand, either end customer demand or, in this case, demand from the downstream step.
- Utiliz: Utilization—A measure of how heavily occupied a piece of equipment is, calculated by dividing takt by effective capacity.
- Uptime/OEE: Overall equipment effectiveness—A holistic measure of equipment performance, considering reliability, yield, and ability to run at design rate.
- Eff capac: Effective capacity—Maximum demonstrated rate times OEE.
- EPEI: Every part every interval—A Lean term for the total cycle over which all product types are made; in a product wheel strategy, this is equivalent to wheel time.

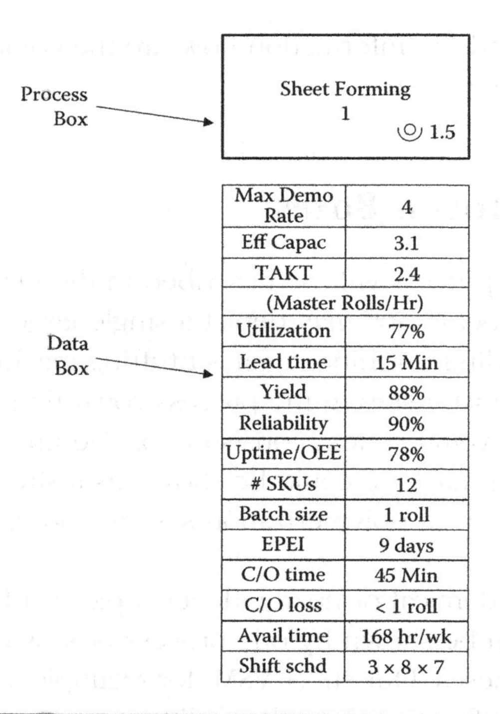

Figure 4.3 Process box and data box—Forming machine 1.

The parameters of primary interest for product wheel design are:

- C/O time: Unless the changeover time is zero or very short, this is a likely candidate for a product wheel.
- C/O loss: The same is even truer for changeover losses.
- # SKUs: The larger the product type count, the more likely that this is a strong wheel candidate, especially when coupled with high C/O time or losses.
- Takt and effective capacity: These are used in the calculations to determine overall wheel time, both in the shortest wheel calculation and in the economic optimum wheel calculation.
- EPEI: The current EPEI represents a workable total production cycle time, without the improvements that product wheels bring. So the wheel time will often be much less than the current EPEI.
- Utiliz: A relatively low utilization, say 70% or less, indicates that we have time for more changeovers and can run a much shorter wheel, if we choose to move in that direction.

Figure 4.4 gives another example, the process box and data box for bonder 2.

Material Flow Icons

There are a number of icons and symbols used to make the material flow portion of the VSM easier to read. Figure 4.5 summarizes those most often used.

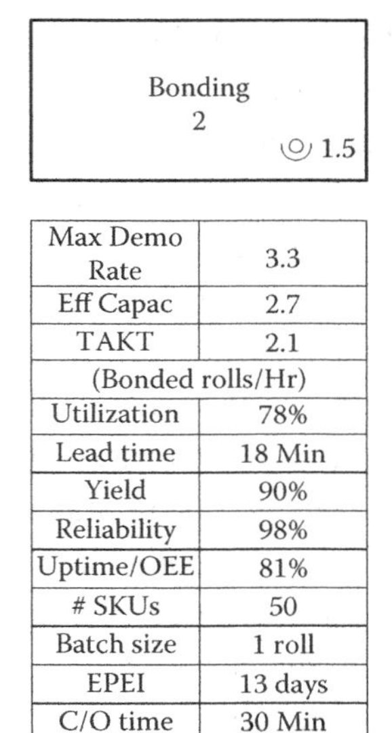

Max Demo Rate	3.3
Eff Capac	2.7
TAKT	2.1
(Bonded rolls/Hr)	
Utilization	78%
Lead time	18 Min
Yield	90%
Reliability	98%
Uptime/OEE	81%
# SKUs	50
Batch size	1 roll
EPEI	13 days
C/O time	30 Min
C/O loss	0
Avail time	168 hr/wk
Shift schd	3 × 8 × 7

Figure 4.4 Process box and data box—Bonder 2.

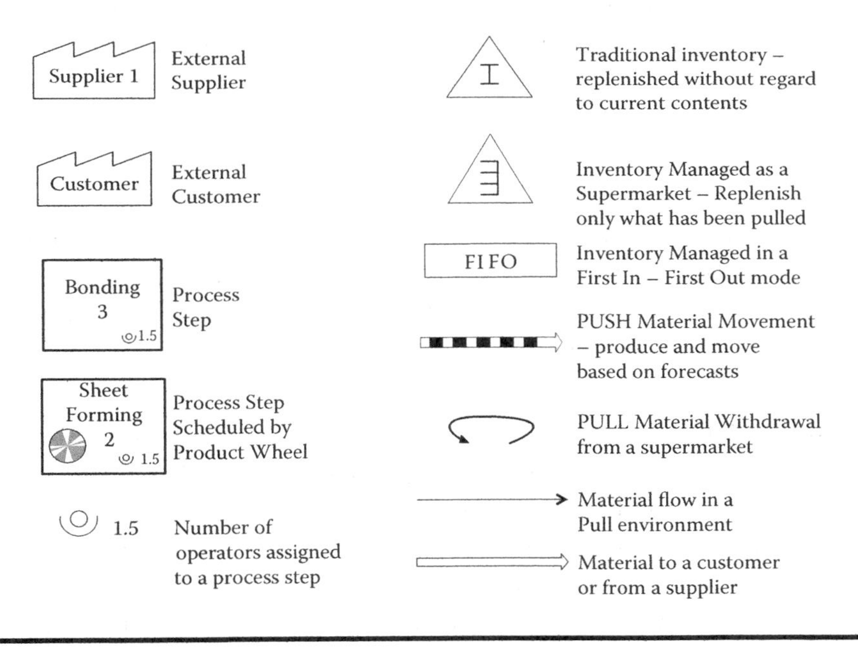

Figure 4.5 Material flow icons.

Inventory Data Boxes

Each inventory, including raw material inventory, work in process (WIP), and finished product inventory is also represented by a data box. Inventory is shown in units of production volume (pounds, gallons, square feet, cubic feet, etc.) and in days of supply. The number of product types stored at that point in the material flow should also be indicated. Figure 4.6 shows the inventories stored after the forming machines, the bonders, and the slitters.

As a rule of thumb, the inventory following any step running a regular production cycle in a make-to-stock environment should be 80 to 100% of the EPEI. If the inventory is 110% of EPEI or greater, then there is a likelihood that we have too much inventory for the current performance. Thus we can probably bring the inventory down simply by understanding how much we need for current conditions, and then reduce it even further with the benefits that product wheel implementation will bring. As an example, from Figure 4.6 we see that the storage between forming and bonding has 11 days of inventory. (Sheet forming has a takt of 9.5 rolls/hour, which is 228 rolls/day. So an inventory of 2500 rolls translates to 2500/228 days, or approximately 11 days.) With forming running a 9-day EPEI, we would expect that stock to be 7 to 9 days, so 11 days looks like too much even for current performance. On the other hand, the inventory after bonding is 12 days, which appears to be appropriate given the 13-day bonder cycle. It should be emphasized that these are just rules of thumb, so the more accurate calculations covered in Appendices A and B should be done before making any firm conclusions. But the map does give you some clues to where an opportunity may lie, to right-size the inventory before applying product wheels, and then with the benefits resulting from wheel implementation.

Information Flow

The information flow portion of the VSM provides some understanding of how customer orders, customer forecasts, back orders, and current inventory data are processed to create the manufacturing schedules now being used. This is useful for product wheel design in several ways:

- It gives an indication of what data are currently available for ongoing product wheel execution, and whether they are available in real time or have delays due to information batching.
- It gives clues to any constraints that may impede scheduling processes.
- Understanding how the schedule generating processes work today gives an indication of what must be changed to incorporate product wheel methods and parameters.

The information flow on the sheet manufacturing process VSM (Figure 4.7) shows that we can get real-time data on customer orders and on inventory status from the database used by the customer service representatives (CSRs). Demand history and forecasts for each specific product should be available from the database used for the sales and operations planning (S&OP) process. The information flow also tells us that each step in the process is scheduled independently, with no apparent coordination, which is probably one reason why

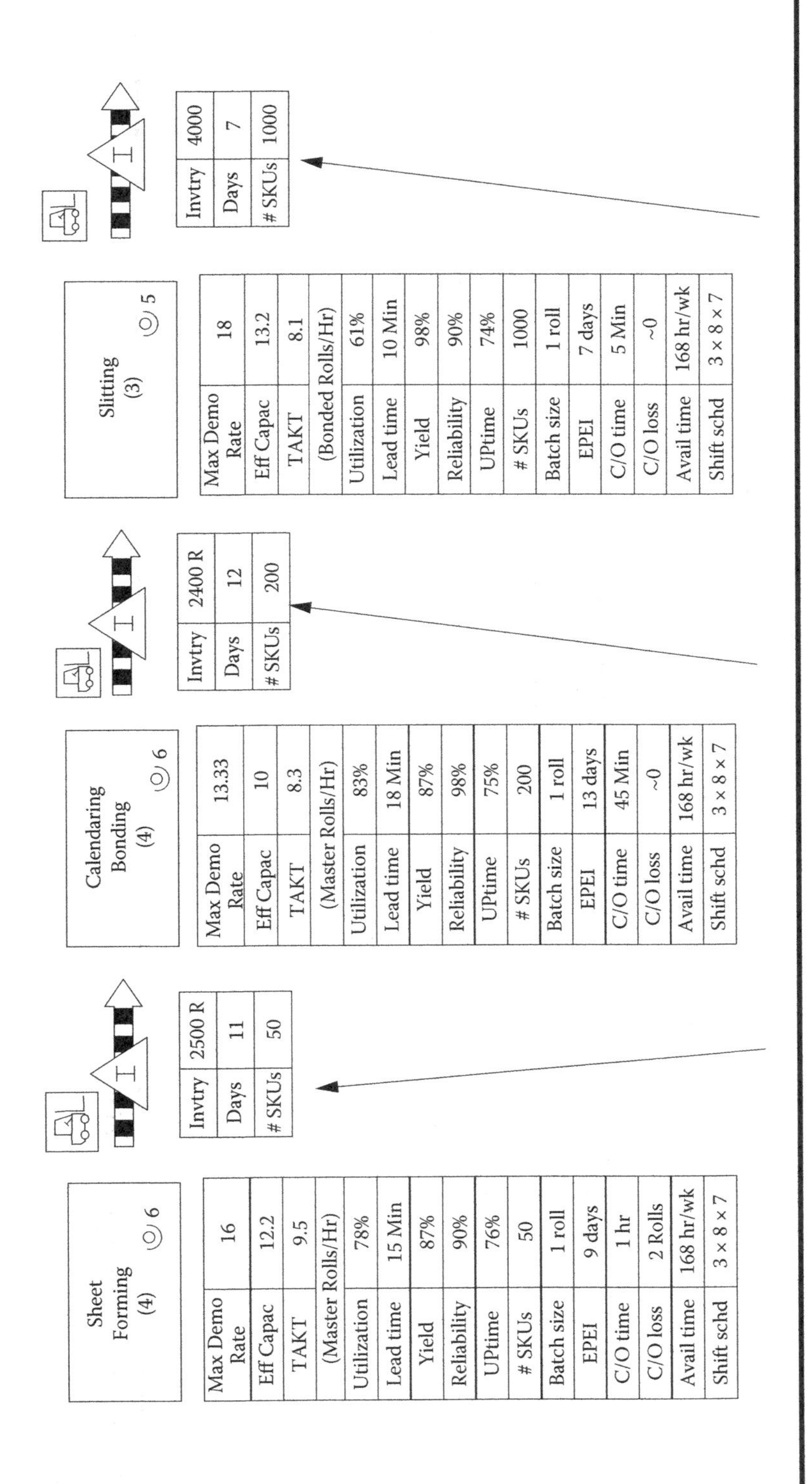

Figure 4.6 Inventory data boxes.

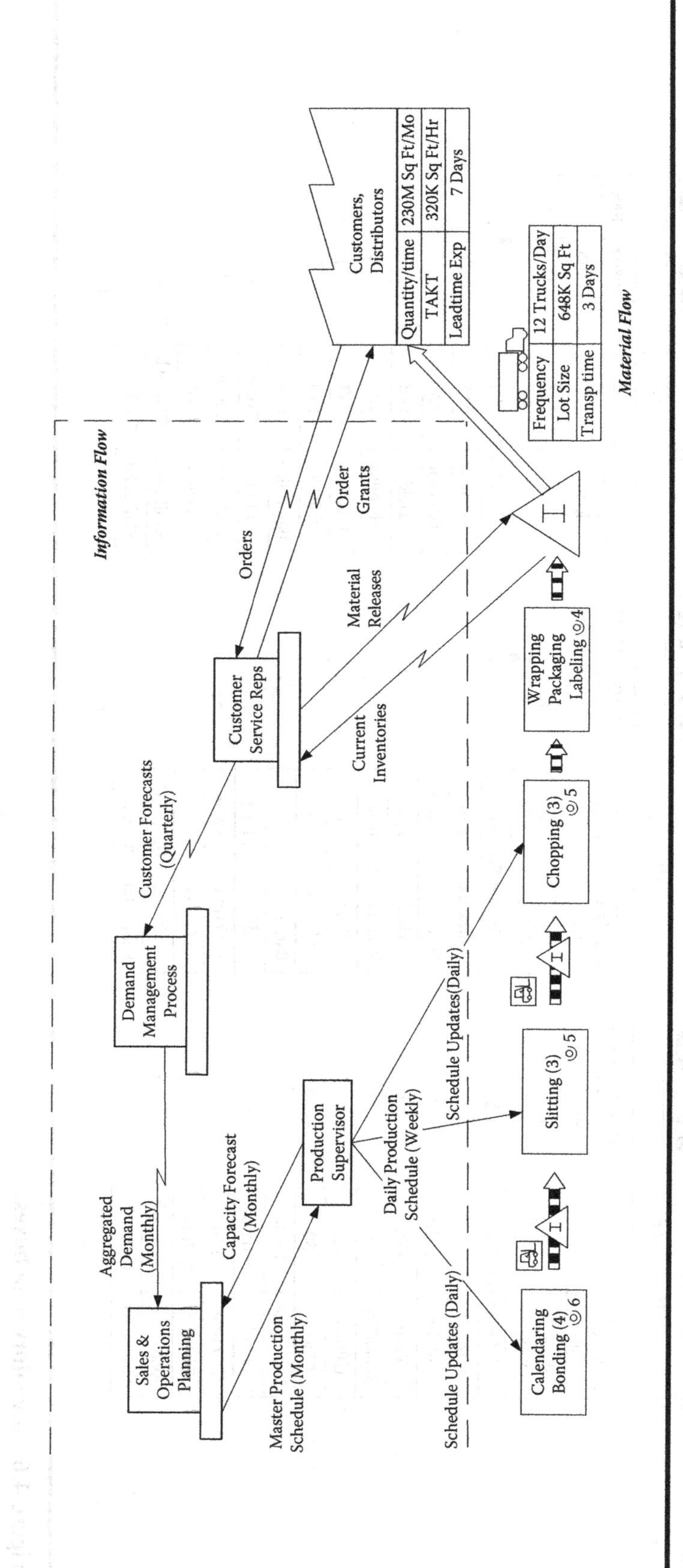

Figure 4.7 Information flow portion of the VSM.

the intermediate inventories are so high. We also see that although schedules are developed weekly, they must be updated daily, probably to accommodate the normal disruptions that can happen in any manufacturing operation. Good product wheel design will result in more stable, more robust schedules, and the ability to take normal disruptions in stride, and therefore dramatically reduce the need for daily modifications.

Summary

Product wheel design requires a thorough understanding of the entire manufacturing process, what causes material to move smoothly through the process steps, and what barriers or limitations exist. Design also requires a significant amount of information about the operating parameters and performance of the specific pieces of equipment. All of this can be seen on a well-constructed value stream map, which is why we emphasize that this should be the first step in the design process.

Chapter 5

Step 2: Decide Where to Use Wheels to Schedule Production

Criteria for Product Wheel Selection

Once we have a map that gives an overview of the entire process, all significant process steps, and the relevant performance data on each step, we can use that to decide which steps should be the focus for product wheels. Any step in the process, any individual piece of equipment, or any entire production line that has appreciable changeover times or losses should be examined as a candidate for its own product wheel. *Appreciable* in this context means any changeover long enough or experiencing enough material loss that it influences the scheduling of the step. In these situations, the product wheel methodology will help to set the optimum campaign length for each product made on that asset. If the time, difficulty, or material loss is sequence dependent, it is even more likely that product wheels will be beneficial, because of the sequence-determining part of the methodology.

Where a production line has continuous flow from one end to the other, such as a salad dressing bottling line (with bottle orienting, filling, capping, labeling, and carton filling), a product wheel will be used to schedule the entire line as a unit. With a process like the sheet goods manufacturing operation described in the last chapter, flow is not continuous or even well synchronized, and each step in the process is separated from the others by in-process inventory. In this case, a product wheel should be applied to each process step that meets the criteria described here.

The key value stream map (VSM) process data box parameters to analyze are:

1. Number of SKUs or product types: A large number of product types means that at the very least, we should be concerned with scheduling strategy and overall production cycle length, and how much of each type we should produce on each cycle.
2. C/O time and C/O losses: High changeover time or changeover losses signify that we must be sure that we are making the various products in the sequence that minimizes these losses. They also identify a need to make sure that the number of changeovers done is optimized in balance with inventory carrying costs.
3. EPEI (every part every interval): A high EPEI is a flag for potential improvement through product wheel application. Large numbers of product types usually lead to high EPEIs, but EPEIs are often higher than needed to process all product types (due to the desire to run long production cycles to avoid changeover losses). The product wheel design methodology can determine if that is the case, and if so, improve it.
4. Steps with a large inventory following that step: Again, this is a flag that overall production cycle lengths may be excessive for the current situation, or that inventory is just not being managed very well. Product wheel design will correct either of these problems.

Secondary indicators:

5. High yield losses: A significant portion of these may be material losses at changeover, another flag that sequencing and scheduling might not be well managed.
6. Low Uptime/OEE (overall equipment effectiveness; see Appendix C for a thorough definition and calculation): Still another flag that sequencing and scheduling may not be well thought out. As total changeover time is a component of OEE, this will show the multiplicative effect of the number of SKUs and changeover time.

Analyze the VSM

Using the above criteria, along with any other operational details of the process that we may understand, we can look at specific steps to see which may be candidates for wheels.

We have made a number of improvements to our sheet process since the VSM shown in the last chapter was created, including reducing changeover times and losses using the single-minute exchange of dies (SMED) methodology (see Appendix D). Cellular flow patterns, shown in Figure 5.1, have been instituted to reduce the number of possible flow paths, which greatly

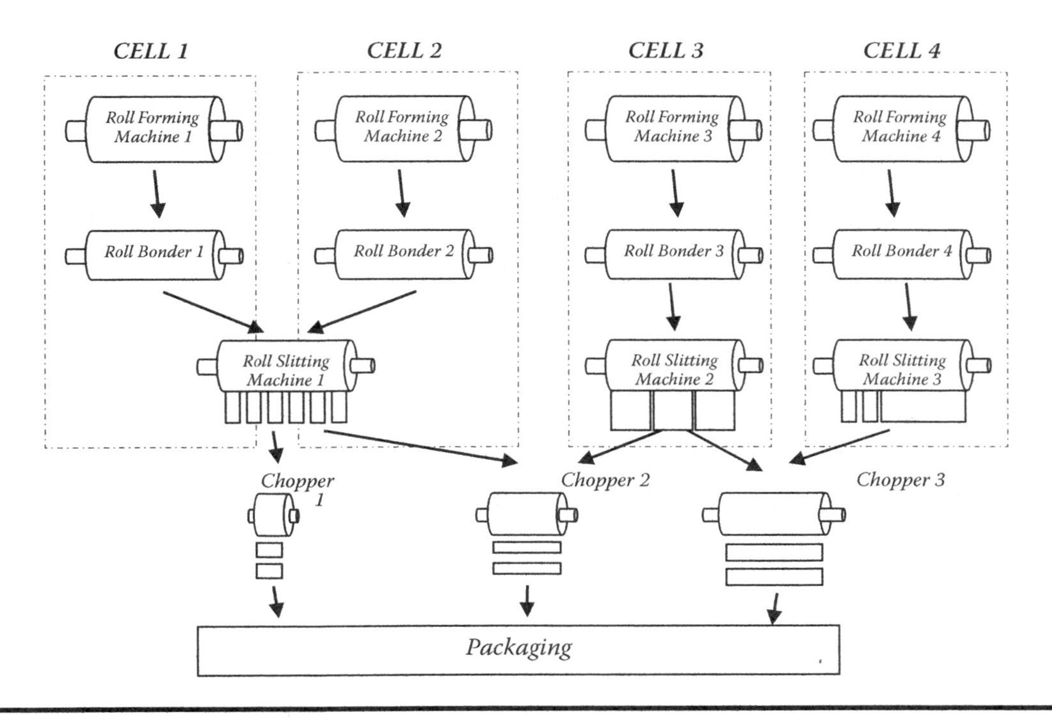

Figure 5.1 Block diagram—Sheet goods process with virtual cells.

simplifies flow coordination and synchronization and, more importantly, product wheel design and execution. Our product lineup has been segmented into families, and each product family is dedicated to a specific cell. So the number of product types that must be included in any product wheel is significantly less than it was before. (Cellular manufacturing is explained in more detail in Appendix F.)

A portion of the resulting VSM illustrating the cellular layout is shown in Figure 5.2

Looking at the VSM for cells 1 and 2, shown in more detail in Figure 5.3, we see that both forming machines meet the criteria for product wheel consideration. They have a number of product types, 12 and 14, have significant changeover losses and times, and currently each run a production cycle of 9 days. Likewise, the bonders should be examined in more detail. Although they have no changeover losses and relatively short changeover times, they process a high number of products, so even short changeovers accumulate to a major time loss. Bonder 2 loses 25 hours every 13 days, or about 8% of the total available time.

Let's examine each specific piece of equipment in more detail.

Forming 1

Looking at the data box for sheet forming machine 1, in Figure 5.4, we see that with the cellular flow pattern, the number of products formed on this machine

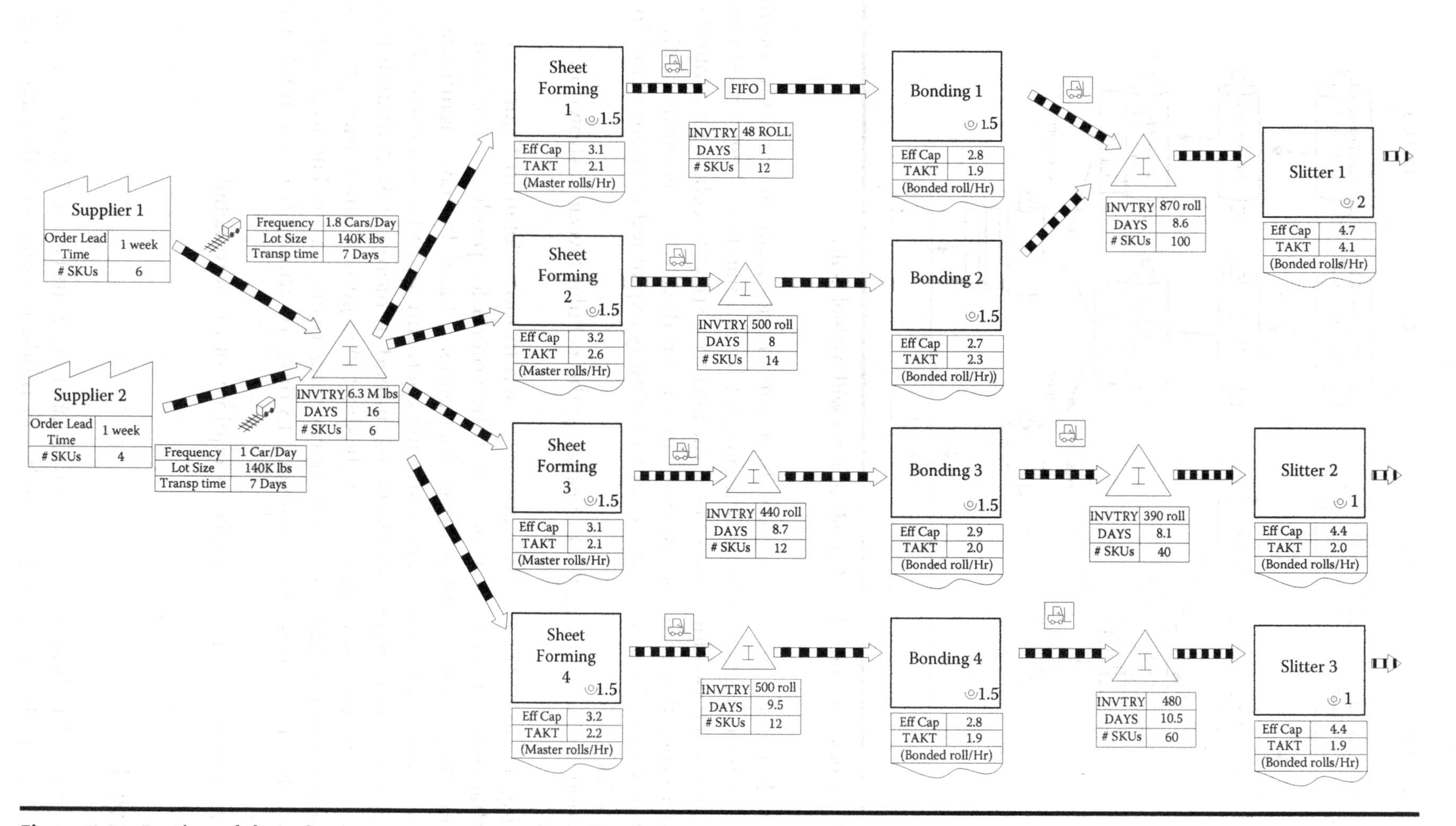

Figure 5.2 Portion of the value stream map after cell application.

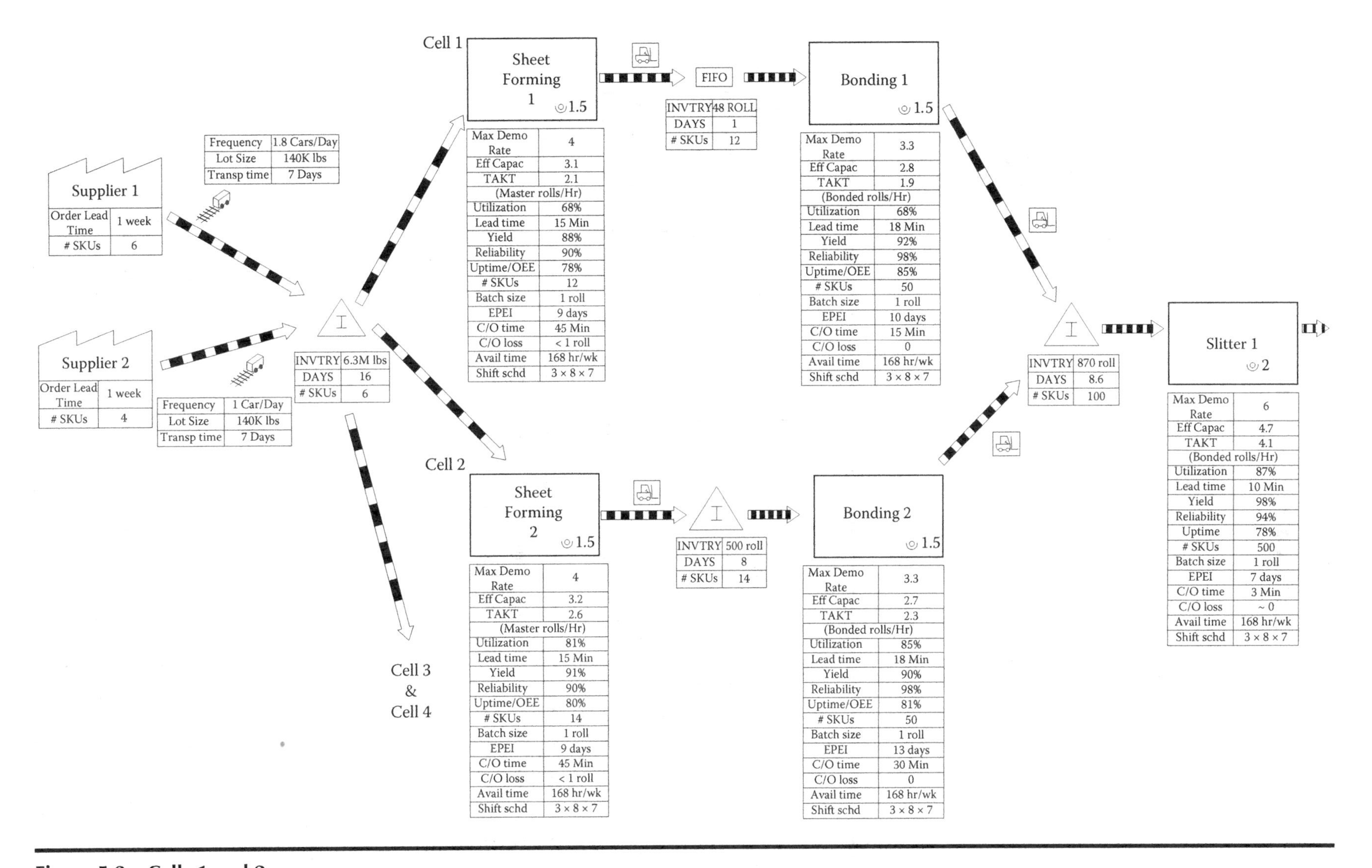

Figure 5.3 Cells 1 and 2.

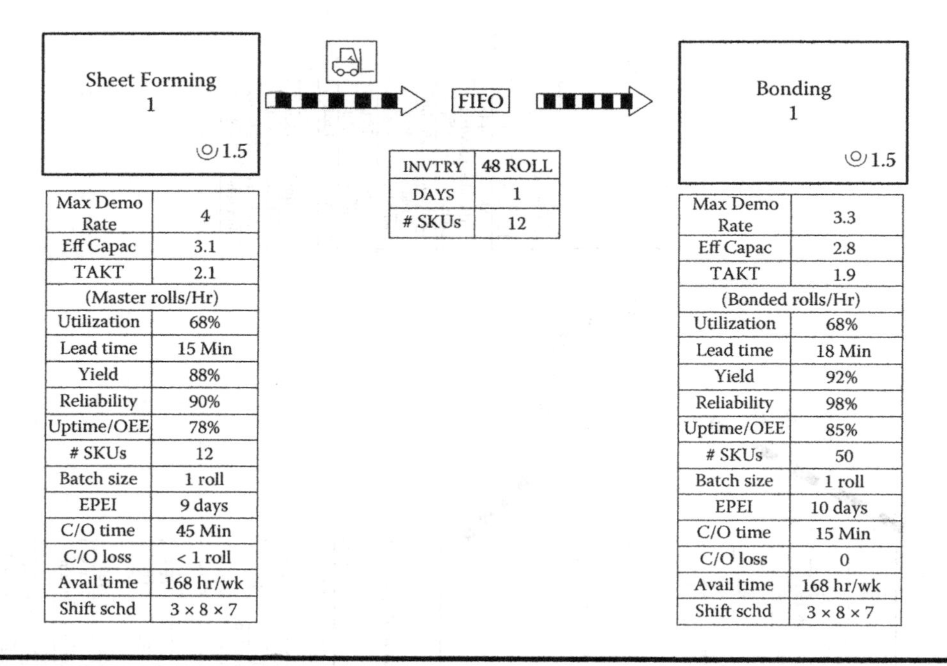

Figure 5.4 Cell 1.

is now 12, changeover time has been reduced from 1 hour to 45 minutes, and the loss on changeovers has dropped from 2 rolls to 1 roll. Even so, we still lose about 12 rolls every 9 days, at significant material cost. We also lose 9 hours of productive time every 9 days. We also know, although we don't see it on the VSM, that changeovers involve changing some combination of four parameters: raw material type, sheet width, basis weight, and winding tension. Each of those has its own set of complications at changeover, so arranging products in the ideal sequence is critical.

So, with an understanding that sequence is very important here, and that cycle length must be optimized to minimize changeover time and losses consistent with good inventory management practice, forming 1 is an excellent candidate for a wheel.

For the same reasons, forming 2 is also very appropriate for product wheel application.

Bonder 2

Figure 5.5 shows the data box for bonder 2. With the cellular flow, there are now 50 products bonded here. There are no materials lost at changeover, and changeover time looks reasonable at 30 minutes. But, as pointed out earlier, that does represent 25 hours of lost production every 13 days, or 8% of total capacity. The only parameter changed on the bonders is the temperature of the heated roll, but it must be changed gradually; the surface of the bonder roll is very precisely machined to be flat within very small tolerances, to provide maximum contact with the sheet, and heating or cooling it rapidly could warp the surface. For that

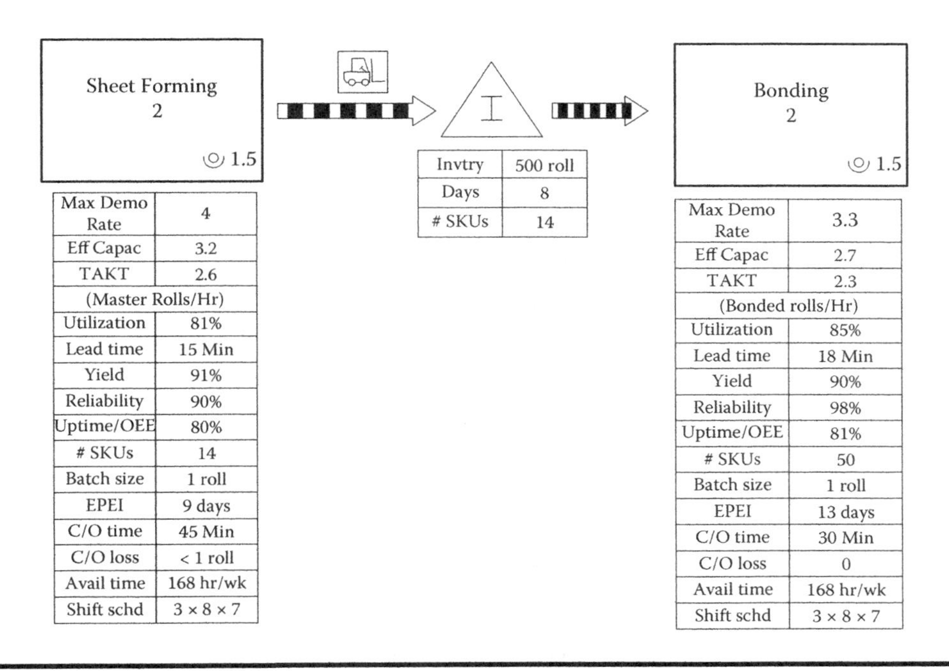

Figure 5.5 Cell 2.

reason it is best to sequence products so that the temperature change at each product change is small. Because some products on bonder 2 are bonded at the same temperatures, sequencing can significantly reduce the total time lost across all changeovers.

Bonder 2 also has a high EPEI, 13 days, which can very likely be reduced with wheel application, so this machine looks like an excellent candidate for a wheel.

Bonder 1

Bonder 1 is a different story. In allocating products to cells as part of the cellular manufacturing design, we were able to find product families to run on cell 1 where the ideal forming sequence naturally groups products by the same bonding temperature; i.e., the optimum forming sequence is the same as the optimum bonding sequence. So we don't need a wheel for bonder 1; we can simply let it follow the sequence coming off forming 1. Thus as products have completed the forming process on forming 1, they flow almost immediately to bonder 1. Some inventory will temporarily build up, simply to buffer the rate differences between the two machines: forming can produce a roll in 15 minutes when running well, where bonding requires 18 minutes per roll. So a small inventory is needed, managed as a FIFO (first in–first out).

As a side note, this highlights one of the advantages of the cellular flow pattern. It very often happens that products can be allocated to cells so that ideal changeover sequences are very similar on the various pieces of equipment, and flow from step to step within a cell can be almost continuous. Hence the emphasis on implementing cellular flow before beginning product wheel design.

Slitter 1 2

Max Demo Rate	6
Eff Capac	4.7
TAKT	4.1
(Bonded Rolls/Hr)	
Utilization	87%
Lead time	10 Min
Yield	98%
Reliability	94%
UPtime	78%
# SKUs	500
Batch size	1 roll
EPEI	7 days
C/O time	3 Min
C/O loss	~0
Avail time	168 hr/wk
Shift schd	3 × 8 × 7

Figure 5.6 Slitter 1 data box.

Slitter 1

We can see from the data box for slitter 1 in Figure 5.6 that there are no changeover losses, and the changeover times are short enough that it might not make sense to campaign products and sequence those campaigns based on slitting pattern. But, as a worst case, if each incoming roll required a different slitting pattern than the previous roll, the 3-minute changeover time across 500 products would occupy 25 hours of the 7-day EPEI, or 15% of the total time available. So even though there is enough spare capacity for all these changeovers, and the worst case is unlikely, there would be benefit from some degree of grouping and campaigning, and a wheel should be given some consideration. This is a judgment call, and an example of the various decisions that must be made in product wheel design when there is no clear, obvious choice.

Summary

The data boxes on the VSM provide almost all the information needed to decide which steps are appropriate candidates for wheels. The key decision criteria are changeover times, changeover losses, and number of SKUs. Additional knowledge about the chemistry and physics involved in each process step is also essential, such as understanding why bonder temperature can't be changed rapidly. That's one reason why we recommend that wheel design be done as a team activity, to benefit from several sources of process knowledge and individual perspectives.

In any process, product wheels can be in series, such as on forming 2 and bonding 2, and in parallel, such as on forming 1 and forming 2. While in this example we have focused on a situation where product wheels are applied to specific pieces of equipment in the process, often a wheel is applied to an entire process line. This is very often the case with packaging lines, which are almost always scheduled as a single integrated system.

Now that we have examined the entire process for product wheel applicability, we need to focus on each selected step, piece of equipment, or process line individually. The remaining steps in wheel design must be done individually for forming 1, forming 2, bonding 2, and slitter 1 (if we choose to use a product wheel there). We will focus on forming 2 for the remainder of the book to show the specific processes, methods, and calculations to be used.

Chapter 6

Step 3: Analyze Products for a Make-to-Order Strategy

The remaining steps in product wheel design are all done on a specific piece of equipment or process line on which we have decided to put a wheel. The next step is to examine the average demand and the demand variability of each product on this wheel, to decide if that product is best made to order or made to stock. (The term *make to order* as used here also includes *finish to order*, a more appropriate term when the process step is beyond the first conversion step and we're not making the order from raw materials, but from some intermediate storage.) The advantages of make to order (MTO) are that no inventory is required beyond the MTO point and there is no uncertainty of demand for that product—you only produce it when you have a firm order. The advantage of make to stock is that it provides a nice level base load and helps with the production leveling requirement.

Demand Volume

We will now turn our attention to forming machine 2 from the process described in the last two chapters. The product grouping done as part of cell design has resulted in 14 products (designated A through N) being assigned to forming 2.

Figure 6.1 lists the average weekly demand and demand variation (σ_D = standard deviation of demand) for each product. It also lists the coefficient of variation (CV), a measure of the relative variability:

$$CV = \sigma_D / D_{avg}$$

Product	*Designation*	*Weekly Demand D (rolls)*	*σ_D (rolls)*	*Coefficient of Variation $CV = \sigma_D/D$*
A	423 J	130	28	0.22
B	403 L	108	18	0.17
C	403 J	68	17	0.25
D	423 L	26	5	0.19
E	426 J	18	3	0.17
F	406 J	15	3	0.20
G	489 J	12	4	0.33
H	406 R	11	3	0.27
I	429 L	10	2	0.20
J	489 R	10	2.7	0.27
K	409 R	9	3.5	0.39
L	409 J	9	3	0.33
M	406 L	8	6	0.75
N	426 R	6	4	0.67

Figure 6.1 Products made on forming machine 2.

When comparing the variabilities of several products, CV is a much better indication of relative variation, where standard deviation is more absolute. For example, product A has the highest absolute variability, 28 rolls, but has a relatively low CV, 0.22, taken as a fraction of the total demand.

Figure 6.2 shows a Pareto chart of the demand data from that table, and illustrates the common trait that a small portion of the total product portfolio has most of the demand volume. (Pareto predicts that 20% of the products have 80% of the demand; the top 3 products on forming 2 are 21% of the product count and comprise 70% of the total demand.)

Demand Variability

Figure 6.3 shows the CV data plotted as a bar chart; it can be seen that products A through F and product I have very stable demand patterns, while products G, H, J, K, and L are somewhat variable, and products M and N are much more variable. It is common that the higher-volume products have more stable demand, while the low-volume products are more sporadic.

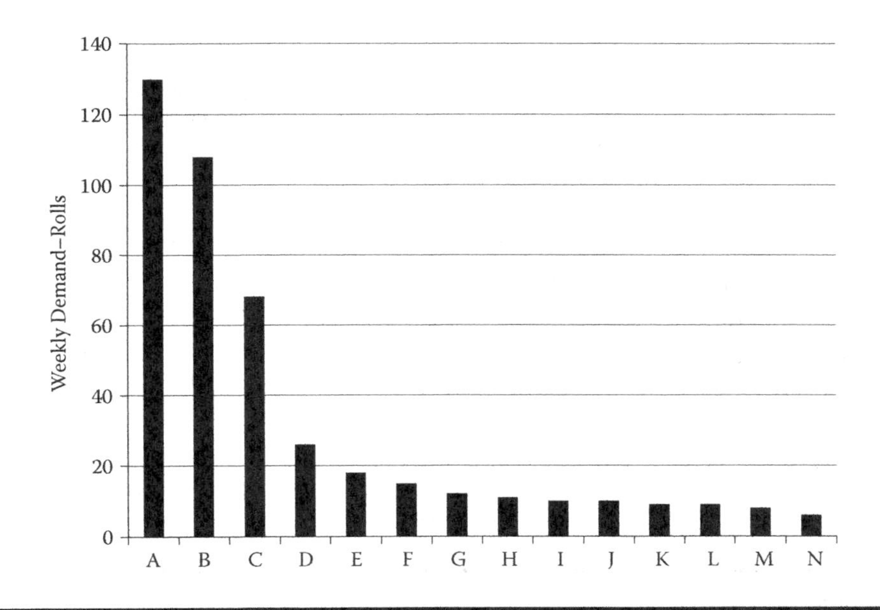

Figure 6.2 Product demand—Weekly volumes.

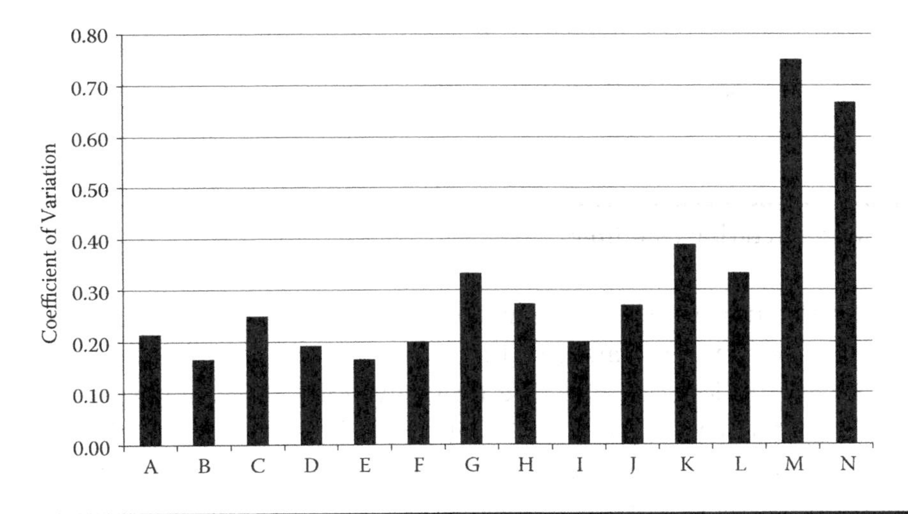

Figure 6.3 Relative variability.

Deciding on the Best Strategy for Each Product

Once each product has been analyzed for average demand and demand variability, the results can be placed on a matrix (Figure 6.4) which can guide decisions on make to order versus make to stock. Products in the lower right-hand quadrant, Quadrant 1, have high demand and very stable demand, and so they should be made to stock. Making the highest-demand products to stock provides a consistent, repeatable base load for the equipment and is what gives product wheels the needed production leveling capability. There is little risk in making these products to stock; they sell on a very regular pattern, so material won't sit in inventory very long.

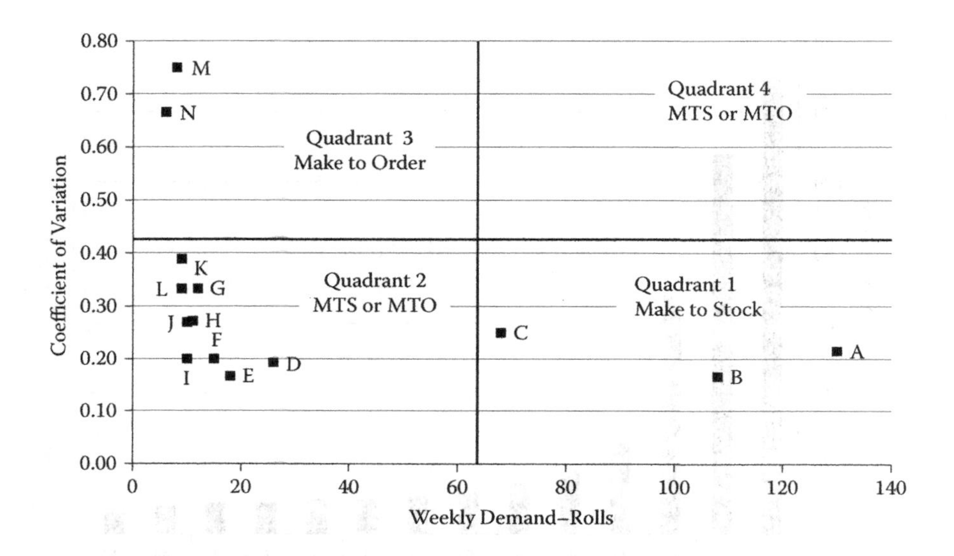

Figure 6.4 Decision matrix.

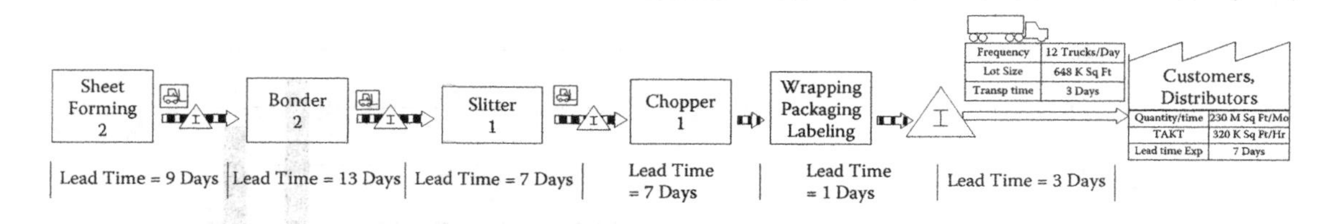

Figure 6.5 Manufacturing lead time.

Products in the upper left quadrant (Quadrant 3) are best made to order: since demand is low and quite variable, if they were made to stock they could sit in inventory for long periods of time. We typically don't know when orders will be placed, and may not know how much will be ordered whenever that is.

There is one firm requirement to be able to make to order: the total manufacturing lead time and shipping time, from the MTO point to arrival at the customer, must be less than the delivery lead time committed to the customer. Figure 6.5 maps this out for forming 2. The total lead time from forming to delivery is 40 days, due largely to the high EPEIs (remember from Chapter 4 that EPEI (every part every interval) is a Lean term for the total cycle time over which all product types are made) in forming, bonding, slitting, and chopping; a product arriving at each of those steps must wait until its turn comes around in the cycle. Since our customer commitment is 7 days, we obviously have an unworkable situation. It is very likely that with product wheel application to forming 2 and bonding 2, those EPEIs, i.e., wheel times, will come down significantly. Even though we may not be putting a wheel on slitter 1, it is very likely that we can reduce the EPEI on slitter 1 and on chopper 1 since changeovers are loss-free and take very little time. But those improvements won't by themselves solve the problem. The real solution will be to negotiate longer lead time commitments with the customers for those low-volume, sporadic products. This is generally an acceptable resolution; customers usually realize that they can't expect immediate delivery for products that they order very occasionally. The dice don't always roll our way, but often a combination of relaxed customer lead time expectations and manufacturing lead time reductions will allow us to make some products to order.

Products in Quadrant 2, with low demand and low variability, could go either way. Making to order lowers our total inventory, while making to stock gives a larger base for production leveling because these products have stable demand patterns. The decision will likely hinge on customers' willingness to accept much longer delivery times for these products, coupled with any process specifics that don't show up in the value stream map (VSM) data.

It is not common for products to fall within quadrant 4, where the demand is high and the variability is also high, unless they are seasonal products, where the demand is very high during the peak season but very low during the slack season, giving rise to the high variability. The best way to handle these products is to make to stock with a high target inventory during the peak season, and an appropriately low target during the slack season. If demand during the slack season is low enough, it might make sense to make to order during the off-season.

Looking at where forming 2 products fall on the matrix, it is clear that A, B, and C should be make to stock, M and N should be make to order, and the rest could go either way. For now, we will plan to make them to stock, because we are not sure we can get customers to relax their lead time requirements for these relatively predictable products.

After the wheel time is set (Chapter 9) and the frequency for lower-volume products has been determined, any borderline-high CV products should have their variability recalculated based on the wheel time and their specific frequency. It is often the case that products have very high variability when measured in very short increments, and that it smoothes out when measured in longer increments. As an extreme example, a product may have very high variability from day to day, but relatively low variability from month to month.

All make-to-order versus make-to-stock decisions can be revisited after wheel time is set, but it is beneficial to get the obvious MTO products off the list to simplify the remaining steps.

Summary

Products that sell in very small volumes and with no predictable pattern should be made only to satisfy specific orders, if that is possible. It doesn't make sense to hold stock for long periods of time when you're not sure if or when the product will be sold. The inventory holding cost frequently eats up all the profit from the sale of those products. From the Pareto principle, these products can comprise as much as 50% of the total product count and 10 to 20% of total finished product inventory when no make-to-order strategy has been implemented.

So before deciding to make these products on some regular frequency on a product wheel, they should be analyzed to see if a MTO strategy is more logical.

Even if we have very short manufacturing lead times and can therefore make everything to order, there is value in making all the higher-volume stable products to stock, to level out production requirements.

Chapter 7

Step 4: Determine the Optimum Sequence

As we've discussed before, long changeover times and changeover material losses generally have more negative influence on production scheduling than any other factors. They tend to drive long production cycles, which in turn lead to high inventories, and more importantly, loss of manufacturing flexibility and agility. And because changeover losses often depend on specific operating characteristics for each product, and can vary depending on which product you're coming from and which product you're going to, it's imperative that you select the optimum sequence to minimize these losses.

So that becomes the next step in wheel design, to determine the optimum sequence in which the various products should be made. In some cases this is very obvious. With products that come in colors, such as house paint, it is generally best to go from very light colors to darker colors in the tinting step, because light colors won't contaminate dark colors as much as the reverse will.

In other cases, even when it is not completely obvious what the best sequence is, people experienced in that operation often believe they understand the best order in which to produce, and often they're correct. But just as often, they're not; they may have a good sequence, close to but not the optimum.

In situations where the best path through the products is not understood, or may be further improved, some analytical view is very helpful.

Changeover Complexity

The thing that complicates understanding the best sequence is that often one or more of several variables must be changed. In the salad dressing and condiment packaging lines found in food plants, changes can involve fluid type, bottle size, bottle labels, cartons, pallet loading pattern, and various combinations of them. The fluid and the label may change, but the bottle size may stay the same, or the fluid may stay the same but be filled into different bottle sizes. Fluid changes will require flushing the lines and filler head, while bottle size changes will require repositioning of conveyor guides and star wheel changes on the rotary filling machine. So there may be a number of combinations of parameters changed, each with its own set of difficulties, times, or material losses

Similarly, bagging lines for lawn care products can experience bag size changes, bag construction changes, label changes, material changes, and combinations of them, each with their specific changeover tasks.

The last step in the production of polyester fibers to be used in 50/50 cotton blends is the cutting of long fibers into ½- to 1½-inch lengths and then baling these short fibers. Changeovers can involve changes to the fiber type being cut, different fiber diameters, and changes to the cut length. Various combinations of these changes require different levels of cleanout and different mechanical adjustments. Trying to find the optimum path through the various combinations by intuition or gut feel rarely gives as good a result as some analytical method will.

It is important to decide whether you want to optimize on changeover time or changeover losses, because they can give different results. In the packaging of motor oil additives, for example, fluid changes require flushing of transfer lines and the resulting loss of valuable material, but can be done fairly quickly. Changing only the bottle size loses no fluid, but takes much longer because all the conveyor guides and other mechanical apparatus must be changed. So optimizing on time would involve cycling through several additive grades going into the same bottle size, and then changing bottle size to sequence through all the grades for the new size. Optimizing on material loss would cause you to run an additive grade through all bottle sizes for that grade, then changing to the next grade. And in the latter situation, there is likely to be an optimum order in which to cycle through the bottle sizes, which must be understood.

Optimizing the Forming 2 Sequence

For forming machine 2 in the sheet goods process, product changeovers can entail changes to the sheet width, winding tension, thickness or basis weight, and raw material polymer type, as shown in Figure 7.1.

The forming machine has automatic roll changing capability on the windup end, so that as a roll has completed the winding operation, it rotates out of the sheet path and an empty core is positioned to receive the sheet. Thus if subsequent rolls are of the same product type, the transfer is instantaneous, and there is no material or time lost. However, if the product type is being changed between rolls, there will be losses.

If only the polymer type changes, there will be transition material in the piping, consisting of the last of polymer 1 mixed with the beginning of polymer 2,

Product	*Designation*	*Sheet Width (ft)*	*Winding Tension (lb)*	*Basis Weight*	*Polymer Type*
A	423 J	12	250	3	J
B	403 L	10	200	3	L
C	403 J	10	200	3	J
D	423 L	12	250	3	L
E	426 J	12	300	6	J
F	406 J	10	240	6	J
G	489 J	8	220	9	J
H	406 R	10	240	6	R
I	429 L	12	320	9	L
J	489 R	8	220	9	R
K	409 R	10	270	9	R
L	409 J	10	270	9	J
M	406 L	10	240	6	L
N	426 R	12	300	6	R

Figure 7.1 Forming 2 changeover parameters.

which represents a time and material loss. Once the sheet consists of only polymer 2, it can take some time for the sheet properties to get within specifications, representing another time and material loss. If, on the other hand, the sheet width is the only parameter being changed, the extrusion dies must be moved in or out across the sheet. The mechanical repositioning is automated, so polymer flow doesn't stop. But it will take time to get sheet properties within the new aim conditions, so again there is time and material loss. Basis weight changes require only a change to the speed of the pump metering the polymer into the extrusion heads, so the primary losses are those while ramping up to goal properties.

Winding tension is changed to alter the mechanical properties of the sheet, primarily tenacity. This involves changing the speed of the take-up roll, and then waiting for properties to stabilize at the new conditions.

We know from experience that the various changes, in order of decreasing difficulty, are polymer type, sheet width, basis weight, and winding tension. One technique to optimize the sequence is to rearrange the columns in the spreadsheet in that order, as shown in Figure 7.2. Then you can sort the spreadsheet with polymer type as the primary sort, width as the secondary sort, then basis weight, and finally tension as the fourth sort parameter. The result is shown in Figure 7.3. We start with polymer J and run all its products, then change to polymer L. Within polymer J, we cycle through all widths, going from narrow to wide, because we know that going wider is a less severe transition than going

Product	*Designation*	*Polymer Type*	*Sheet Width (ft)*	*Basis Weight*	*Winding Tension (lb)*
A	423 J	J	12	3	250
B	403 L	L	10	3	200
C	403 J	J	10	3	200
D	423 L	L	12	3	250
E	426 J	J	12	6	300
F	406 J	J	10	6	240
G	489 J	J	8	9	220
H	406 R	R	10	6	240
I	429 L	L	12	9	320
J	489 R	R	8	9	220
K	409 R	R	10	9	270
L	409 J	J	10	9	270
M	406 L	L	10	6	240
N	426 R	R	12	6	300

Figure 7.2 Parameters sorted by difficulty.

from wide to narrow. And within the 10-foot widths, we cycle through basis weight going from thin to thick, because experience teaches us that makes transitions easier.

(We could of course sort directly on the table shown in Figure 7.1; the column sort was done to illustrate the logic being employed, and to make the final pattern shown in Figure 7.3 more visible.)

In cases where the parameters affecting changeover are more complex and interrelated, a from/to changeover matrix like the one shown in Figure 7.4 can be helpful. In even more complex situations, there are operations research tools that can be applied to find the optimum sequence.

Optimizing the Sequence in Complex Situations

In some cases, the optimum sequence may not be obvious, and not clearly seen through spreadsheet tables. In these situations, more rigorous mathematical techniques can provide guidance toward the optimum solution.

As an example of a complex combination of changeovers, a line packaging various grades of brake fluids and transmission fluids has several things that may or may not have to be done on a product change. If the fluid is changed, the lines may or may not have to be flushed, depending on the degree of difference

Product	*Designation*	*Polymer Type*	*Sheet Width (ft)*	*Basis Weight*	*Winding Tension (lb)*
G	489 J	J	8	9	220
C	403 J	J	10	3	200
F	406 J	J	10	6	240
L	409 J	J	10	9	270
A	423 J	J	12	3	250
E	426 J	J	12	6	300
B	403 L	L	10	3	200
M	406 L	L	10	6	240
D	423 L	L	12	3	250
I	429 L	L	12	9	320
J	489 R	R	8	9	220
H	406 R	R	10	6	240
K	409 R	R	10	9	270
N	426 R	R	12	6	300

Figure 7.3 Products sorted by difficulty.

in the fluids. For large differences in fluid grade, filters may require cleaning. If the bottle size is being changed, it might be a diameter change, a height change, or both. If only height is being changed, perhaps only the sealer and capper will require repositioning. If only the diameter is changing, then the conveyor guides, rotary filler table, and capper will likely have to be changed. If the new bottle is different in height and diameter, all of those will have to be adjusted. Because the same fluid may be sold in several bottle sizes, and the same bottle size can be used for several fluid grades, there are a wide variety of combinations of things that must be cleaned or adjusted on a product changeover.

Let's look at a very simple version of that situation to explain some of the more powerful approaches that can be taken. Imagine Figure 7.4, but instead of easy, moderate, and difficult, each cell had a changeover time in it. It is possible for the table to be symmetric or asymmetric. If the table is symmetric, the changeover time from product A to product B is the same as the changeover time from product B to product A. This is not always the case. A might contaminate B, but B does not contaminate A. In this case there would be additional time needed to make sure that there are no traces of A left in the equipment, so it would take longer to change over from A to B than from B to A.

If only a few products are to be made, it is easy to go through all of the possible sequences and determine the shortest, as shown in Figure 7.5. However, the number of possible sequences is $N!$ (factorial), with N as the number of products

	From													
To	*F1*	*F2*	*F3*	*F4*	*J1*	*J2*	*J3*	*J4*	*R1*	*R2*	*R3*	*R4*	*R5*	*R6*
F1		E	E	E	M	M	M	M	M	D	D	D	D	D
F2	E		E	E	M	M	M	M	M	D	D	D	D	D
F3	E	E		E	M	M	M	M	M	D	D	D	D	D
F4	E	E	E		M	M	M	M	M	D	D	D	D	D
J1	M	M	M	M		E	E	E	D	D	D	D	D	D
J2	M	M	M	M	E		E	E	D	D	D	D	D	D
J3	M	M	M	M	E	E		E	D	D	D	D	D	D
J4	M	M	M	M	E	E	E		D	D	D	D	D	D
R1	M	M	M	M	D	D	D	D		E	E	E	E	E
R2	D	D	D	D	D	D	D	D	E		E	E	E	E
R3	D	D	D	D	D	D	D	D	E	E		E	E	E
R4	D	D	D	D	D	D	D	D	E	E	E		E	E
R5	D	D	D	D	D	D	D	D	E	E	E	E		E
R6	D	D	D	D	D	D	D	D	E	E	E	E	E	

Figure 7.4 Product from/to matrix. E, easy; M, moderate; D, difficult.

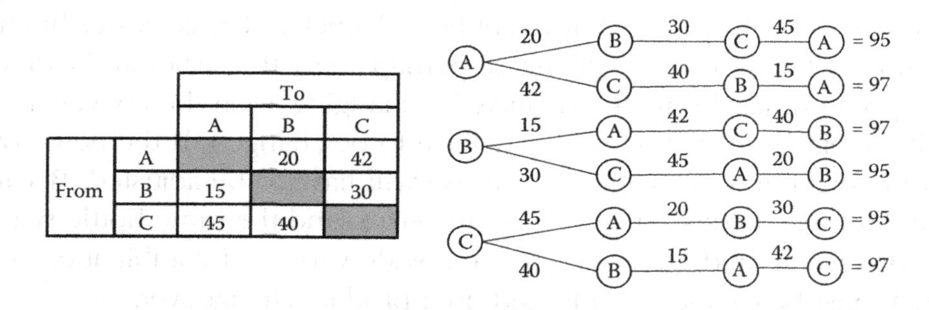

Figure 7.5 Analyzing sequence combinations.

that must be made. If N is more than 3 or 4, this method could take a while. With 3 products, there are only 6 possible sequence options, but with 10, the possibilities jump to 3.6 million, and with 14 products, the number of sequence options becomes 87 billion!

So with more than a very few products, we obviously need a more powerful approach. The traveling salesman problem (TSP), a classic operations research problem, provides one such approach. It involves figuring out the shortest overall route a salesman visiting N cities can take. The routes are called tours. In the case of product wheels, the salesman is the process equipment, and each city is a different product. The distances between the cities are changeover times.

The TSP has been studied for over 75 years, and various methods can be used to solve it, depending on how large *N* is. There are world records for solving large TSPs. There are whole books dedicated to different ways to solve them. There is even an iPad app! Developing a complete solution is beyond the scope of this book, but we will discuss the basic approach.

The simple three product cases above can be set up as an optimization to illustrate the technique. Let *tij* be the changeover time from *i* to *j*, and *xij* = 1 if you make the changeover from *i* to *j* (that is, if you produce product *i* immediately before product *j*) and *xij* = 0 otherwise. So the total changeover time is sum(*tij***xij*) for all *i* and *j*. The objective is to minimize this sum.

$$x_{12}t_{12} + x_{21}t_{21} + x_{13}t_{13} + x_{31}t_{31} + x_{23}t_{23} + x_{32}t_{32}$$

Now we must consider some constraints. We must make all of the products, and make them each only once, so sum(*xij*) = 1 for all *i* and sum(*xij*) = 1 for all *j*.

$$x_{1j} + x_{2j} + x_{3j} = 1$$

$$x_{i1} + x_{i2} + x_{i3} = 1$$

Also, you need to make sure that the result doesn't have you going from one product to the same product, so *x*11, *x*22, and *x*33 must be 0. With more than 3 products, the solution might result in a collection of subtours, which are unconnected cycles. We must eliminate those solutions, which makes the problem even more complicated. This can be approached using linear programming, but that's beyond the scope of this book.

In addition to linear programming, there are other approaches that can be explored, including numerous heuristics for determining a solution. The nearest-neighbor algorithm starts with a random city (product), goes on to the closest city (product with the shortest changeover time), and continues visiting the closest unvisited city (unmade product) until all cities are visited (all products are made). This method can be improved by calculating the total changeover time for each starting product, and choosing the lowest time. This will not give you the optimal solution, but you only have to go through *N* permutations.

These methods are not as simple or straightforward as the methods described earlier in this chapter, but in some cases the economic implications of finding the very best solution warrant the extra effort.

Summary

Determining the optimum sequence to cycle through the products made on any production line or major piece of equipment is one of the most important aspects of wheel design. Find the best sequence, and wheels will run at their best; settle for a less optimum sequence, and wheel performance will suffer.

Operations people usually think they've already got the best sequence, and often they're right, but the implications of sequencing are so great that it is very beneficial to use some analytical method to illustrate the patterns so that the best arrangement can be found.

Spreadsheets listing products and changeover parameters, from/to diagrams, and other visual techniques can generally lead to an appropriate sequence. When things are more complex, other decision-making tools are available, but require familiarity with advanced operations research methods.

As a final thought, applying cellular manufacturing before product wheels, as we have done here, and making a carefully thought out assignment of products to cells, will usually lead to products with similar processing requirements being grouped together, and thus simplify changeovers and make the sequencing problem much more straightforward. Cellular manufacturing is explained in detail in Appendix F.

Chapter 8

Step 5: Analyze the Factors Influencing Overall Wheel Time

Once the sequence of products on the wheel has been established, the next step is to decide how long a single cycle of the wheel should be. Deciding on wheel time is one of the most important decisions in the entire process, in that many other steps are built on that, and more importantly, it determines all the economic factors that product wheels can affect.

There are at least five factors that must be considered in determining wheel time, and perhaps more depending on any unique constraints found in a specific situation. Some of these won't apply in every situation, and no one of them always gives the optimum answer, but all can provide useful information for making the wheel time determination.

- Time available for changeovers—the shortest wheel possible
- Economic order quantity (EOQ)
- Short-term demand variability
- Minimum practical lot size
- Shelf life

These will generally give different answers, and no one of them should be considered the best answer, but they each give useful perspectives that should be considered in making the wheel timing decision.

Time Available for Changeovers—The Shortest Wheel Possible

One very useful piece of information when deciding on the overall wheel time is the fastest wheel that could practically be run. You may not want to run the very shortest, fastest wheel possible, but it is helpful to know what it would be, as it sets a lower limit on wheel time.

The available time model considers a fixed period of time, perhaps a week. It then computes the amount of time required to produce the full customer demand for that period and subtracts that from the total available time, i.e., the planned operating time for that period. The difference is the amount of time that could be available for changeovers. The total changeover time for one cycle of the wheel is calculated by adding up all of the individual product changeovers. That sum is then divided into the total time available for changeovers to indicate how many cycles could be completed per week. The wheel time will then be the number of hours per week divided by the number of cycles per week.

Total available time – Total production time = Time available for changeovers

Time available for changeovers/Sum of changeover times =
Wheel cycles per period

Total available time/Wheel cycles per period = Wheel time

With a takt of 2.6 rolls/hour, forming machine 2 must produce 436 rolls per 168-hour week to meet customer demand. At the effective capacity of 3.2 rolls/hour, that will require 136 hours of actual production, leaving 32 hours (168 – 136) for changeovers. With the virtual cell lineup in place, forming 2 produces 14 product types, requiring an average of 45 minutes each, so a complete set of changeovers consumes 10½ hours.

This is shown graphically in Figure 8.1. As 32 hours is available for changeovers each week, and one complete set takes 10½ hours, we have time to do about three complete sets of changeovers and thus run three wheel cycles per week. The total available time of 168 hours divided by 3 gives us a wheel time of 56 hours. This represents the shortest wheel time possible under the stated conditions; it is not to suggest that it is the best wheel time, just the minimum possible.

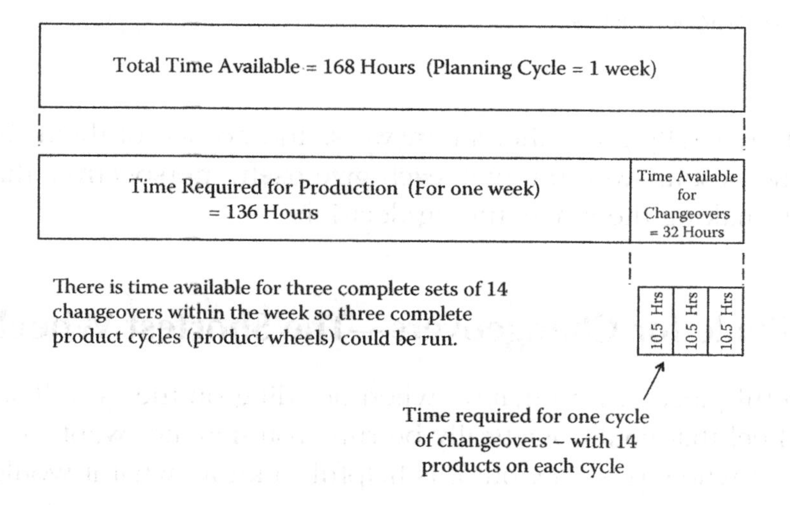

Figure 8.1 Time available for changeovers.

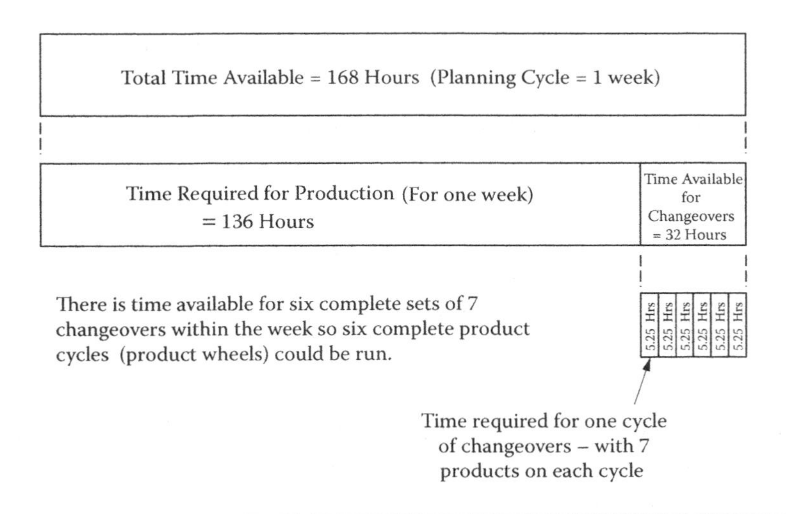

Figure 8.2 Time available for changeovers—Seven products on each cycle.

This assumes that every product is made on every wheel cycle, which will generally not be the case. The wheel time is usually controlled by the highest-volume products, and only high-volume products are made on every cycle; lower-volume products are made less frequently, so that there is a reasonable amount produced each time the product is made. Looking at product demand from Figure 6.2, it appears that A, B, and C have sufficient demand to be made every cycle, and the rest made perhaps every second or third cycle. So it is reasonable to expect that each wheel cycle will produce A, B, and C, and perhaps four of the lower-demand products, so we will make seven products per cycle, and require seven changeovers, for a total of 5.25 hours per set. On that basis, the 32 hours available for changeovers would allow six sets of changeovers, with a wheel time of 28 hours. Figure 8.2 illustrates that situation.

This is not to imply that a 28-hour wheel is what we want, but simply to set the bar, that the wheel must be at least 28 hours long. If it is strategically important to the business that we have the shortest possible manufacturing lead time, this would be a good choice. If minimizing manufacturing cost is most strategically important, then the alternative described in the next section would be a better choice.

Finding the Most Economic Wheel Time

Another very important factor to consider when deciding on the fundamental wheel time is the balance between the high inventory costs associated with long wheels and the high changeover costs associated with short wheels. A classic way to find the optimum balance between these two conflicting factors is the economic order quantity (EOQ) calculation, also known as economic production quantity (EPQ). The available time calculation just described provides the shortest wheel possible under the current conditions, but not necessarily the best wheel possible. By taking economic factors into account, the EOQ model gives another useful perspective.

The specific equation for the production quantity that results in the lowest total changeover plus inventory cost is

$$\mathrm{EOQ} = \sqrt{\frac{2 \times \mathrm{COC} \times D}{V \times r \times \left(1 - \frac{D}{PR}\right)}}$$

where COC = changeover cost, D = demand per time period, V = unit cost of the material, r = percent carrying cost of inventory per time period, and PR = production rate in units per time period.

(Note that the time period must be the same for all factors. If demand is in rolls per week, r must be the annual percentage divided by 52.)

(A simpler equation for EOQ is sometimes presented:

$$\mathrm{EOQ} = \sqrt{\frac{2 \times \mathrm{COC} \times D}{V \times r}}$$

The equation being recommended here accounts for the fact that some of the material produced is being consumed during its spoke, so that the inventory at the end of the spoke is slightly less than the sum of the amount produced and the starting inventory.)

The EOQ calculation returns a quantity, the production quantity that theoretically results in the lowest total changeover and inventory cost. To convert it to a wheel time, you must divide the quantity resulting from EOQ by the demand over some period of time. So if the EOQ result for product A on forming 2 is 152 rolls, and we divide that by the product A demand of 130 rolls/week, we get a recommended wheel time of 1.17 weeks, or 8.2 days.

The EOQ calculation must be run separately for each product on the wheel, which will likely give a different result for each product. Reconciling these differences will be covered later.

Figure 8.3 shows the EOQ concept graphically. As wheel length increases, the inventory cost can be seen to rise, the changeover cost declines due to the decreasing number of changeovers, and total cost drops and then rises. The total cost has its minimum at the wheel time where the two cost components intersect.

It must be noted that the EOQ calculation is only an approximation, and that it takes into account the factors that typically are most significant, but ignores some factors usually having lower impact. The inventory included, for example, is only the cycle stock required, that is, inventory to satisfy average demand between production spokes for a material. It ignores safety stock, inventory needed to protect against normal random variation in supply and demand.

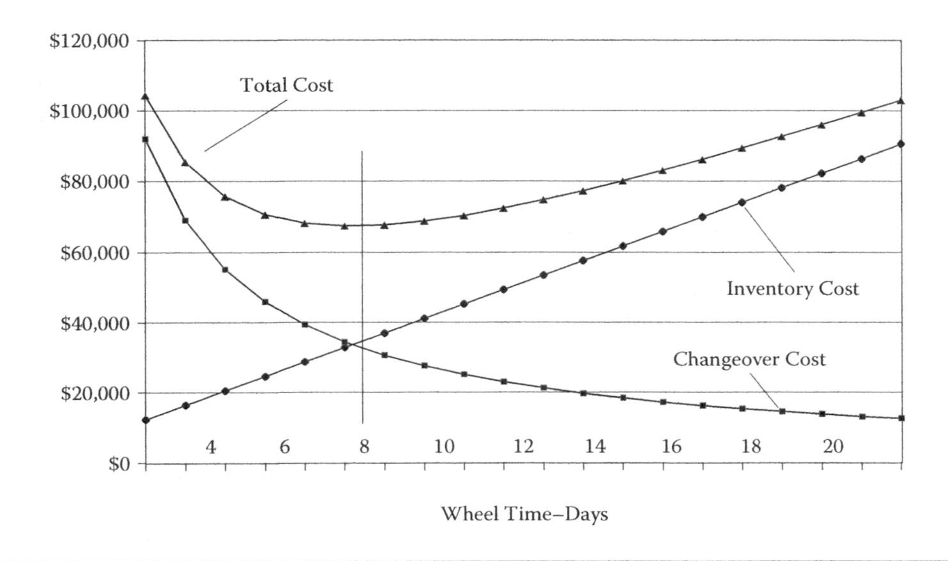

Figure 8.3 Economic order quantity curves for product A.

It should be obvious that for EOQ to be meaningful, the changeover cost must be known to a reasonable degree of accuracy. In our sheet forming process the changeover cost varies with the product to be run and the product that was just run. Any disruption to the process, whether it be a polymer change, a width change, a basis weight change, a tension change, or any combination of them, causes the product properties to go outside the specification limits, so some time is required to bring properties back within tolerance. In the past, up to two full rolls might be formed before acceptable properties were reached; today, through the application of SMED techniques, on-line instrumentation to give immediate quality feedback, and better process control algorithms, the product reaches acceptable limits well within the first roll formed. Since these improvements are relatively recent, we don't have exact data on the new changeover costs. Therefore we have estimated the cost of each changeover, as shown in column 5 of Figure 8.4. The fact that changeover costs are not always precisely known is another reason why EOQ is an approximation.

An interesting property of the EOQ curve is that total cost is very flat in the region of the minimum. The consequence of this is that selecting a wheel time somewhat greater or less than the optimum will be very nearly as good. Selecting a wheel length 20% lower than the optimum will increase total cost by only 3%; selecting a 30% longer wheel time also increases total cost by only 3%.

These two factors, the flatness of the curve and the fact that it is only an approximation, reinforce that the EOQ result should be considered a guide, not a precise number to be followed exactly. Practical considerations must also be considered. Many operations function more smoothly if the wheel time is some integer number of days or weeks. If the higher-volume products suggest a wheel time of 6 to 9 days, for example, and the wheel time is therefore set to 7 days, the fact that the cycle repeats weekly may make it easier to follow. Everyone involved, schedulers, planners, operators, maintenance,

1	2	3	4	5	6	7	8	9	10
Product	*Designation*	*Weekly Demand D (rolls)*	*Product Cost per Roll*	*Changeover Cost*	*Inventory Carrying Cost*	*EOQ (rolls)*	*Optimum Frequency (days)*	*Selected Frequency (days)*	*Demand per N Cycles (cycle stock)*
G	489 J	12	$1920	$640	27%	39.7	23.2	21	36
C	403 J	68	$1800	$540	27%	94.8	9.8	7	68
F	406 J	15	$2100	$525	27%	38.5	18.0	14	30
L	409 J	9	$2400	$680	27%	31.6	24.6	21	27
A	423 J	130	$2160	$756	27%	151.9	8.2	7	130
E	426 J	18	$2520	$630	27%	42.3	16.5	14	36
B	403 L	108	$1800	$630	27%	134.9	8.7	7	108
M	406 L	8	$2100	$525	27%	28.0	24.5	MTO	N/A
D	423 L	26	$2160	$756	27%	60.7	16.3	14	52
I	429 L	10	$2880	$720	27%	31.3	21.9	21	30
J	489 R	10	$1920	$480	27%	31.3	21.9	21	30
H	406 R	11	$2100	$580	27%	34.6	22.0	21	33
K	409 R	9	$2400	$600	27%	29.7	23.1	21	27
N	426 R	6	$2520	$630	27%	24.2	28.2	MTO	N/A

Figure 8.4 Results of EOQ calculation showing one possible choice.

and test labs, can get into a regular pattern. If there is excess capacity, the wheel will end before the next cycle must begin, and if this occurs at regular, repeatable times, preventative maintenance can be scheduled in advance for the idle periods.

EOQ will give a different production quantity and ideal cycle time for each product because several of the factors in the equation are product specific. Thus setting an overall wheel time requires some adjustment and minor compromising to find the best balance. Figure 8.4 shows the results of the EOQ calculation for paper forming machine 2: the optimum production quantity and the frequency, i.e., wheel time, suggested for each of the eight products. (Note that the rows have been rearranged to reflect the sequence selected in Chapter 7.) We can see from column 8 that the optimum cycle times range from 8.2 to 24.6 days, excluding the products that are to be made to order. One reasonable fit is to base the wheel on a 7-day cycle (column 9), and make products A, B, and C on every 7-day wheel cycle. Products D, E, and F would be made on every second cycle, every 14 days, while G, H, I, J, K, and L would be made every third cycle, every 21 days. Remembering the flatness of the EOQ curve, these are all reasonably good fits to the recommended optimum. This is just one possibility; there may be other equally good fits, which will be explored in the next chapter, on making the final decision.

In this case, although there were differences in changeover costs, all were in the same ballpark. In some situations, products fall within families, where the changeover within a family is low in cost, but changeover from family to family is very expensive. In these cases, an adjustment to the basic EOQ calculation, developed by Silver and Peterson (1985) in *Decision Systems for Inventory Management and Production Planning*, can optimize total inventory and changeover costs accounting for the simpler within-family transitions.

If there are no costs associated with changeovers, if changeover cost (COC) is zero, the EOQ calculation is meaningless; it gives a zero production quantity and an infinitely short wheel. In these cases, the available time–shortest wheel model may give the most appropriate wheel time. For example, bonder 2 has no losses on any changeover; the only drawback to changeovers is the 30 minutes it takes to change roll temperature. With no changeover costs, there is no economic penalty to running very short wheels, so running the shortest wheel that allows for the required number of the 30-minute changeovers may be the best strategy for bonder 2.

If the process is extremely capacity limited, the above is not true. In those cases, time lost represents capacity lost, and thus revenue lost. Therefore the margin lost on lost sales should be included in the COC used in the EOQ calculation.

Leveling Out Short-Term Demand Variability

For some products, variability can be quite high when measured in short time increments, but smoothed out over longer intervals. If, for example, you have

Week	Demand	Week	Demand	Week	Demand	Week	Demand
1	78	14	98	27	70	40	694
2	102	15	125	28	796	41	122
3	101	16	845	29	102	42	67
4	741	17	71	30	81	43	98
5	107	18	122	31	70	44	643
6	75	19	118	32	891	45	97
7	108	20	960	33	135	46	127
8	940	21	136	34	78	47	135
9	65	22	79	35	82	48	658
10	92	23	107	36	904	49	139
11	95	24	638	37	136	50	87
12	470	25	95	38	64	51	87
13	94	26	109	39	102	52	608

Figure 8.5 Demand variability data—Breakfast cereal product.

products where the customer ordering practice is to place most orders near the end of a week, say on Thursdays and Fridays, variability measured on a daily basis will be extremely high, but variability measured in weekly increments may be quite stable. For those products, it may make sense to produce at lower frequency than EOQ would suggest.

Let's consider a relatively low-volume breakfast cereal product, with an average demand of 250 cases per week. The higher-volume products made on this line have caused us to set the wheel time at 1 week, and EOQ for this product recommends a frequency of 2 weeks. But the two largest customers generally order once every 4 weeks, toward the end of each 4-week period. Figure 8.5 gives the demand history for this product for the past 52 weeks, including orders from all customers. Figure 8.6 shows this graphically, and illustrates the preponderance of orders at 4-week intervals. If we follow EOQ, and produce at a frequency of two wheel cycles, 2 weeks, we would be trying to produce to the pattern shown in Figure 8.7. This pattern has a standard deviation of 350 cases, and a CV of 0.67. If instead of following EOQ, we choose instead to go to a 4-week frequency and produce on every fourth wheel cycle, we have the pattern shown in Figure 8.8, which is clearly much smoother. It has a standard deviation of 152 cases and a CV of 0.14, a small fraction of the CV for the 2-week case.

If the recurring pattern was not recognized, and the product was set to be made on a 2-week frequency, the safety stock required would be very high, 578 cases, to cover the high short-term variability, assuming that 95% is the desired

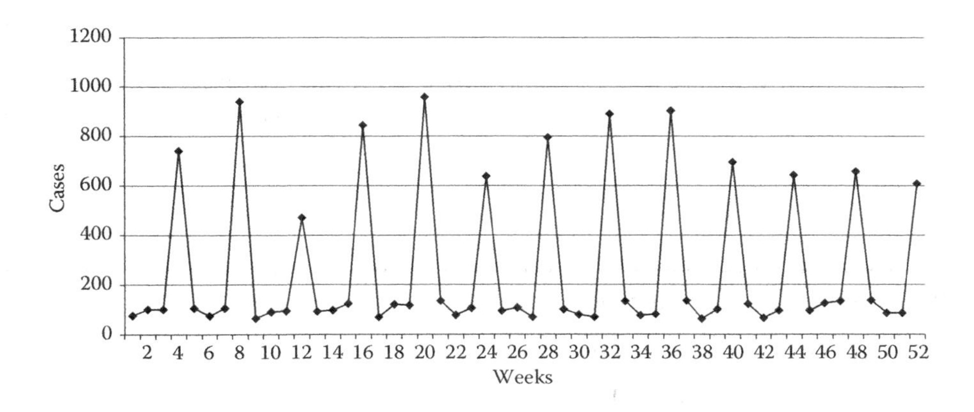

Figure 8.6 Demand variability profile—Breakfast cereal product.

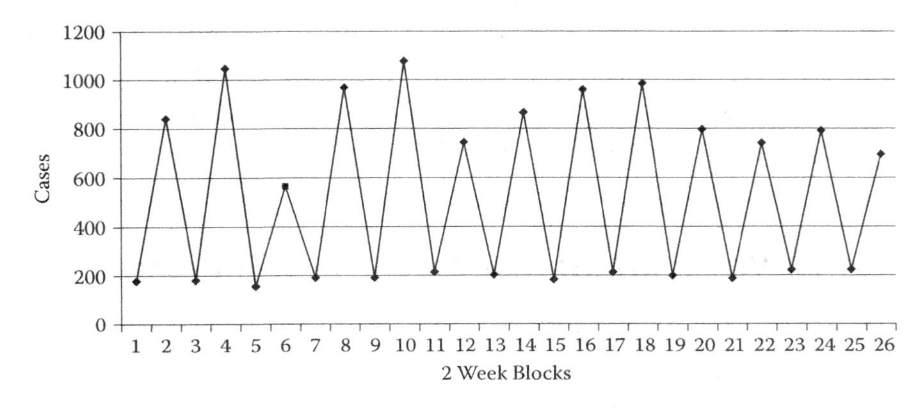

Figure 8.7 Breakfast cereal demand grouped into 2-week increments.

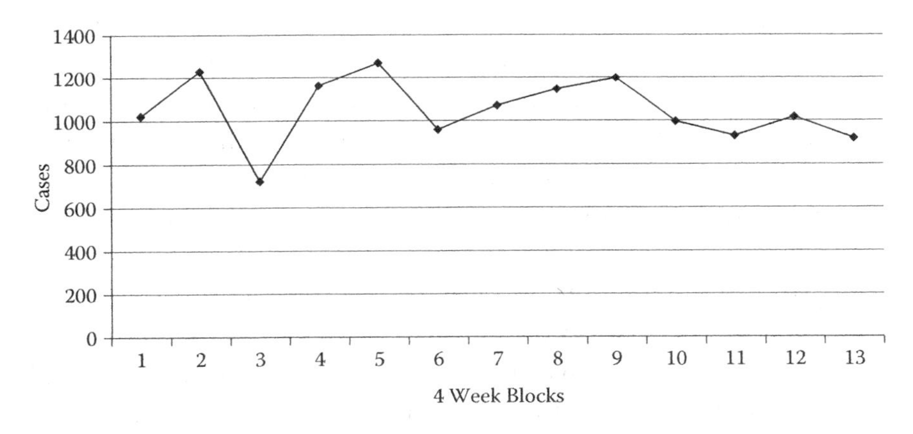

Figure 8.8 Breakfast cereal demand grouped into 4-week increments.

cycle service level. (Details of safety stock calculations are given in Appendix B.) Peak inventory would be 1078 cases, and average inventory would be approximately 828 cases. If, on the other hand, the recurring pattern is recognized and a 4-week frequency is selected, the lower variability calls for only 251 cases of safety stock, and an average inventory of 751 cases. Thus the inventory required to support the longer cycle is actually less because of the dramatically reduced variability, as detailed in Figure 8.9.

Frequency	*Cycle Stock (cases)*	*Safety Stock (cases)*	*Peak Inventory (cases)*	*Average Inventory (cases)*
2 weeks	500	578	1078	828
4 weeks	1000	251	1251	751

Figure 8.9 Inventory requirements for 2- and 4-week frequencies.

Because we have recognized that the demand peak occurs on a relatively repeatable 4-week period, it would be beneficial to slot this product on the wheel so that it is produced just before the peak hits.

In any case where EOQ suggests a frequency that results in a high CV, the demand pattern should be examined to see if there is a recurring phenomenon such that a lower frequency would significantly reduce CV. If we ignore this, the high variability will cause us to carry large safety stock inventory, and yet it won't be enough to protect us from stock-outs when the peaks occur.

An Additional Word about Standard Deviation and CV

The breakfast cereal situation just described is only relevant if the demand is, in statistical terms, not normally distributed.

If the demand for a product is normally distributed, i.e., its histogram follows a normal bell-shaped curve, the variability measured by standard deviation will exhibit the opposite behavior than that seen with the cases of cereal, and will increase if the time increments used in the measurement increase. For example, the variability of the monthly demand will be greater than the variability in weekly demand, and the variability in annual demand will be even greater. That is true of the variability measured by standard deviation, which is an absolute measure of variability. The relative variability as measured by CV will, however, decrease as the time increments of measure increase.

As an example, if the variability in demand for product A is normally distributed with a standard deviation of 28 rolls per week, the standard deviation measured in 4-week increments would be about 56 rolls, and the annual variation would be about 200 rolls. (Standard deviation tends to increase by the square root of the time increment increase.) But because the average demand on a 4-week basis is 520 rolls, the CV drops to 0.11. And with an annual demand of 6760 rolls, the CV measured on an annual basis would be 0.03. Figure 8.10 illustrates this. Thus as the time increments increase, the absolute variability increases while the relative variability decreases.

The reason for bringing up this perhaps obscure statistical point is to prevent the assumption that the treatment of the breakfast cereal product can be broadly applied. For many products, the demand variability is relatively close to being normally distributed, so the behavior illustrated in Figure 8.10 can be expected.

Measurement Increment	*Average Demand (rolls)*	*Standard Deviation (rolls)*	*CV*
Weeks	130	28	0.22
4 weeks	520	56	0.11
Annual	6760	200	0.03

Figure 8.10 Product A—Variability as a function of measurement increments.

And because safety stock requirements depend on variability as measured by standard deviation, the safety stock required for a product made on every fourth cycle of a 1-week wheel will be about twice the safety stock required by a similar product made weekly.

However, with the breakfast cereal example described above, the weekly variability is very definitely not normally distributed. In fact, it behaves the opposite of a normal distribution: the standard deviation decreases as the time increments used to group the measurements increase. That's why the safety stock required decreased as the time covered by a spoke increased.

Thus it would be a mistake to conclude that the way we dealt with the breakfast cereal product is broadly applicable. But the situation does occur often enough that the data must be analyzed to understand if the demand pattern is normal or nonnormal to determine what course of action is appropriate.

Making Practical Lot Sizes of Each Material

In some cases, there is a minimum lot size below which it is not practical to produce. On a high-speed line packaging automotive fluids, for example, the line was capable of packaging four full pallets of cases of fluid in less than an hour. With all of the adjustments that must be made to the filler head table and to the conveyor guides and labelers, it didn't make sense to make a change any more frequently than every 2 hours, so a minimum lot size of eight pallets was considered reasonable. On forming machine 2 in our sheet goods process, considering that we lose 45 minutes and some part of one roll on every product change, we may decide that it makes no sense to run any less that an 8-hour campaign. So the minimum lot size would be set at, say, 26 rolls.

If the EOQ method is being followed, minimum lot size is generally not an issue. EOQ will usually suggest campaign sizes longer than the practical minimum; it will propose a campaign length where the amount produced justifies the cost of the changeover. But with very expensive materials, or with equipment that can be changed over at low cost, that won't necessarily be true. With very expensive products, the inventory carrying cost will drive EOQ to very short runs. Inexpensive changeovers, even if they take a lot of time, will also move the EOQ breakpoint toward short runs. Thus either or both of these situations may push the EOQ recommendation below a practical minimum lot size. The underlying reason is that EOQ considers the cost of changeovers but not the time lost in changeovers. (The exception is cases where the line is running at capacity, so time lost is revenue lost, and the lost revenue is included in changeover cost.)

The spinning of Kevlar fibers, for example, requires expensive raw materials and very costly processing, so carrying a lot of inventory is prohibitively expensive. Changeovers can take time, but don't involve a lot of variable cost, so the EOQ calculation will produce a recommendation that is impractically short. Therefore a minimum lot size should be set that accounts for the time lost on changeover, while keeping the inventory cost within reasonable limits.

In these situations, the minimum lot size consideration usually trumps the EOQ suggestion. There can also be cases where EOQ makes very reasonable recommendations for the high-volume products, but drives some low-volume products to impractically short runs. Those low-volume products should be scheduled at a wheel frequency lower than the EOQ guidance, to allow a reasonable minimum production quantity. Referring back to Figure 8.4, we can see that product L is almost in this situation; its suggested run is 27 rolls, very close to the 26-roll minimum we set above. And had we not decided to make product N to order, it would have been below the threshold, at an EOQ recommendation of 24 rolls.

Protecting Shelf Life

In some cases, shelf life specifications constrain wheel time. This is often seen in the food industry, where shelf life limitations come up against high changeover costs.

Food companies must produce an ever-expanding array of product varieties to remain competitive today, with fat-free, low-calorie, gluten-free, and other variations added to the standard product lineup. Then there is the issue of products containing allergens such as peanuts coupled with the need to make allergen-free versions. This high variety doesn't allow dedicated lines for families of products, so extensive cleanups are required on changeover. Not only do the old materials have to be completely flushed out, but the lines and other equipment must be thoroughly sanitized, and then tested for cleanliness and freedom from even extremely small amounts of contaminants. Thus there is the cost of cleaning fluids and laboratory time in addition to the line time lost and the cost of product lost.

This tends to drive the EOQ calculation toward very long campaigns; 8 to 12 weeks is not unusual. But shelf life limitations won't allow that. Many food products have a stated shelf life of 12 months. The retailer wants at least 9 months of that left when he puts the product out on his shelf; consumers are becoming ever more conscious of expiration dates. Distributors typically want 2 months to get

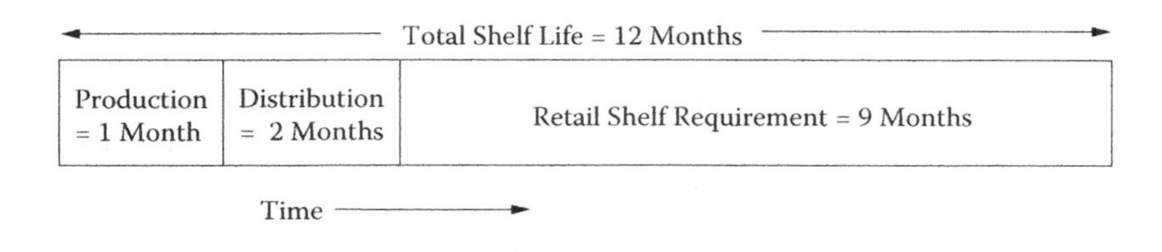

Figure 8.11 Allocation of product shelf life.

the material from the producer, through one or more levels of distribution center, and to the retailer. That leaves 1 month for the producer, as shown in Figure 8.11.

Thus the EOQ analysis points to a wheel time of 2 months or more, while shelf life considerations constrain the wheel to a month or less. Anytime shelf life issues come up against any of the other factors, shelf life wins, and so we are forced to a much shorter wheel than we would prefer.

The situation can possibly be helped by using SMED and other techniques to try to simplify and shorten changeover time and cost, by making sure that the laboratory is extremely responsive on any cleanliness testing that must be done, and by insuring that clean-in-place (CIP) technology is being used to the fullest. Work can also be done to extend the shelf life, but this requires changing customer and retailer attitudes and expectations, which may prove difficult, in addition to whatever it takes to actually extend the useful life of the product.

The point is that anyone dealing with shelf life limitations must be sure they are considered in setting wheel time.

Making to Stock Using a Trigger Point

If there are low-demand products that can't be made to order, and so must be made to stock, it is often better not to make them to a specific wheel frequency, but instead to make them when a trigger point is hit. In some cases this can significantly reduce inventory.

The inventory profile for a trigger point product was shown in Figure 2.3.

As an example, if we have a 1-week wheel, and a product R, which the EOQ analysis says should be produced every eighth cycle, we could make that product on exactly every eight wheel cycles. If we plan to make it only every 8 weeks, we must carry enough safety stock to cover whatever variability we expect to see over 8-week cycles.

On the other hand, we can let the frequency vary, and make the product only when the inventory drops below some threshold, called a trigger point. If product R can be made on any wheel cycle, if there is sufficient capacity to fit it in on any cycle, and if every cycle has a sequence where R can be inserted without unreasonable changeover difficulty, then we can be confident that R can be made at any time needed with a 1-week lead time. So the trigger point is the expected demand during that 1-week lead time, plus enough safety stock to protect against the demand variability expected during the 1-week lead time. Thus the safety stock needs to protect against only 1 week of variability instead of 8, so it drops to about 1/3 of the safety stock that would have been required.

On sheet forming machine 2, we had decided that product M should be made only to order. If market conditions won't allow that, if customers require delivery of this within the normal lead times, we must have the required inventory in stock. The EOQ analysis suggested that M be made every 28 days, every fourth cycle of the 1-week wheel. If we do that, we must carry enough safety stock to cover 4 weeks of variability. If we make it only when the inventory drops below a trigger point, and can fit it in on any wheel cycle, we need only 1 week of safety stock coverage. Since the time at risk has dropped by a factor of four, the safety stock can be cut by the square root of that, i.e., in half. The trigger point would be the expected demand during the 1-week lead time, 8 rolls, plus the safety stock, 10 rolls (see Chapter 12 for a discussion on calculating safety stock), for a total of 18 rolls.

If a trigger point strategy is being used, it is important to restrict it to the very low-volume products only. If trigger points are used for more than a few products, we begin to lose the structure, predictability, and production leveling that wheels provide.

Summary

There are several questions to be considered when deciding on the total wheel cycle time. What is the fastest wheel we could run? What is the most cost-effective wheel? Is there short-term variability that must be considered? What is the minimum campaign size we can practically run? Are there shelf life considerations? And there may be others related to your specific products. The answers may all point in somewhat different directions, so the best choice may not be completely clear from the numerical answers. Like several other decisions to be made in product wheel design where there is no single clear-cut answer, judgment must be exercised. But looking at overall wheel time from several points of view provides very helpful perspectives and useful guidance, which coupled with your knowledge of the specific details of your process will lead you to the right selection.

The next chapter will explain how the various trade-offs can be dealt with after all of these factors have been analyzed, along with any other considerations that are relevant in your situation, in order to make the final wheel time determination.

Chapter 9

Step 6: Put It All Together—Determine Overall Wheel Time and Wheel Frequency for Each Product

Now that we understand most of the factors that can impact the selection of wheel time, we can look at how they apply to forming machine 2 and its product lineup. We'll start with economic order quantity (EOQ), because that generally has the most influence on wheel time.

EOQ—The Most Economic Wheel Time

The primary benefit of EOQ is that it gives the best economic balance between changeover costs and inventory carrying costs, and therefore the lowest total cost. For that reason, it is the tool of choice in cases where minimizing manufacturing cost is of utmost importance. It has the added benefit of providing a rational, logical method to decide how frequently the lower-volume products should be made.

We saw in the previous chapter (Figure 8.4) that a 7-day wheel would be a good fit for the frequencies suggested for the various products made on forming machine 2. Figure 9.1 shows that 8 and 9 days are also reasonable possibilities. Running an 8-day wheel, with products made every 8, 16, or 24 days, actually provides a slightly closer fit to the suggested optimum for 9 of the 12 make-to-stock products. A 9-day wheel also looks very good: it provides the closest fit for 3 of the 12 products.

Product	*Designation*	*Optimum Frequency (days)*	*Frequency (days) 7-Day Wheel*	*Frequency (days) 8-Day Wheel*	*Frequency (days) 9-Day Wheel*
G	489 J	23.2	21	24	27
C	403 J	9.8	7	8	9
F	406 J	18.0	14	16	18
L	409 J	24.6	21	24	27
A	423 J	8.2	7	8	9
E	426 J	16.5	14	16	18
B	403 L	8.7	7	8	9
M	406 L	24.5	Make to order		
D	423 L	16.3	14	16	18
I	429 L	21.9	21	24	27
J	489 R	21.9	21	24	27
H	406 R	22.0	21	24	27
K	409 R	23.1	21	24	27
N	426 R	28.2	Make to order		

Figure 9.1 Three possibilities for wheel time. Shaded cells = best fit for this product.

Given that we have at least three reasonable choices, we will select the 7-day alternative. The key reasons are:

- The weekly regularity is an advantage to operations. Having the wheel repeat on a 7-day cycle makes planning and scheduling much easier. We will have about 24 hours of process improvement time available each week for preventive maintenance, continuous improvement kaizen events, operator training, new product test runs, and other valuable tasks. So if we start the wheel on Thursday morning, for example, operations can plan on having the full 24-hour day every Wednesday, at the end of the weekly cycle, for PIT time use.
- We will also be running product wheels on forming machines 1, 3, and 4, and on bonders 2, 3, and 4. Running wheels on a weekly cycle allows us to stagger the wheels so that only one wheel has its PIT time fall on any given day, thus leveling out the need for support facilities and resources.
- A 7-day wheel gives a slightly shorter manufacturing lead time than the other two alternatives, which, although slight, can be important in some market environments.

- It is reasonably close to the economic optimum—Figure 9.2 shows a detailed comparison of the 7- and 8-day wheels, illustrating how the inventory costs and changeover costs balance out for each product. Figure 9.3 gives a summary cost comparison for 7, 8, and 9 days, and also what the economics would be if each product were made at the optimum frequency. Of course, it would be impossible to construct a wheel that includes the optimum frequency for each specific product: 8 days, 9 days, 10 days, 16 days, 22 days, etc., as we must fit all products to a fundamental cycle or integer multiples of it. But even though that combination can't work, the calculations are useful to provide a basis of comparison for the three choices that will work. We can see that the totals are all within a couple of percent of each other, which reinforces the observation that the curve is very flat in the optimal region. And because of the approximations inherent in the EOQ calculations, the margin of error is much greater than 2%, so the totals can be considered to be equal.

Thus for the purposes of forming machine 2, we will base our wheel design on EOQ, and choose 7 days as the overall wheel time. Eight or 9 days would have been equally good choices if the weekly regularity were not important to operations, and if the shorter lead time offered no advantage.

The Shortest Wheel Possible

Running the shortest wheel possible gives the shortest process lead time, which may be vitally important for products in very dynamic markets. It is also preferable from a Lean perspective, in that it makes the smallest lot sizes and moves closer to one-piece flow (or in this case, one-roll flow). The analysis in the previous chapter indicated that we could run a wheel as short as 28 hours. That was based on the premise that we would be making only about 7 of the 14 products on each cycle, and the EOQ table in Figure 9.1 bears that out.

Although we could run a wheel that short, it allows no margin for error, and so a wheel time somewhat longer than the shortest possible is usually more practical. Running at something longer than the minimum wheel gives a buffer for significant failures and other unexpected events, and provides time for maintenance and continuous improvement activities. A 2-day wheel is still very close to the minimum, and provides about 13 hours each week for PIT time activities, so it would be a reasonable compromise if we want to use the shortest wheel strategy.

The drawback to a cycle that short is that making changeovers that frequently would heavily utilize maintenance mechanics, test lab facilities, and other resources involved in changeovers well beyond their current capacity. This was one of the factors that led to the selection of the 7-day wheel time.

Product	Designation	7-Day Wheel				8-Day Wheel			
		Frequency (days)	Inventory Carrying Cost	Changeover Cost	Total Cost	Frequency (days)	Inventory Carrying Cost	Changeover Cost	Total Cost
G	489 J	21	$9124	$11,124	$20,248	24	$10,427	$9733	$20,161
C	403 J	7	$14,443	$28,157	$42,600	8	$16,507	$24,638	$41,144
F	406 J	14	$8269	$13,688	$21,956	16	$9450	$11,977	$21,427
L	409 J	21	$8602	$11,819	$20,421	24	$9831	$10,342	$20,173
A	423 J	7	$28,782	$39,420	$68,202	8	$32,894	$34,493	$67,386
E	426 J	14	$11,839	$16,425	$28,264	16	$13,530	$14,372	$27,902
B	403 L	7	$20,995	$32,850	$53,845	8	$23,995	$28,744	$52,738
M	406 L	Make to order				Make to order			
D	423 L	14	$14,433	$19,710	$34,143	16	$16,495	$17,246	$33,741
I	429 L	21	$11,448	$12,514	$23,962	24	$13,083	$10,950	$24,033
J	489 R	21	$7632	$8343	$15,975	24	$8722	$7300	$16,022
H	406 R	21	$9165	$10,081	$19,246	24	$10,474	$8821	$19,295
K	409 R	21	$8602	$10,429	$19,031	24	$9831	$9125	$18,956
N	426 R	Make to order				Make to order			
					$367,894				$362,979

Figure 9.2 Detailed cost comparison—7-day and 8-day wheels.

Product	Designation	Total Cost 7-Day Wheel	Total Cost 8-Day Wheel	Total Cost 9-Day Wheel	Total Cost Optimum Frequency
G	489 J	$20,248	$20,161	$20,383	$20,149
C	403 J	$42,600	$41,144	$40,470	$40,343
F	406 J	$21,956	$21,427	$21,277	$21,277
L	409 J	$20,421	$20,173	$20,253	$20,169
A	423 J	$68,202	$67,386	$67,665	$67,386
E	426 J	$28,264	$27,902	$27,997	$27,902
B	403 L	$53,845	$52,738	$52,544	$52,544
M	406 L	Make to order			
D	423 L	$34,143	$33,741	$33,887	$33,741
I	429 L	$23,962	$24,033	$24,452	$23,939
J	489 R	$15,975	$16,022	$16,301	$15,959
H	406 R	$19,246	$19,295	$19,624	$19,224
K	409 R	$19,031	$18,956	$19,171	$18,943
N	426 R	Make to order			
		$367,894	$362,979	$364,024	$361,577

Figure 9.3 Cost summary comparison—7-, 8-, and 9-day wheels and optimum frequency.

Short-Term Demand Variability

Looking back at Figure 6.1, we see that most of the forming 2 products have very reasonable CVs, ranging from 0.17 to 0.39, on a 1-week basis. Because all products except A, B, and C will likely be made on a 2-week or longer frequency, the CVs should decrease. The only exceptions are the two products to be made to order, M and N, where demand variability is not a concern. Thus short-term demand variability won't be a factor.

Minimum Practical Lot Size

With the frequencies we have chosen, all planned production quantities exceed the minimum lot size we set at 26 rolls, although K and L come close, at 27 rolls each, as shown in column 10 of Figure 8.4.

Shelf Life

If there are shelf life requirements, they must be accommodated, taking precedence over the other considerations. But all of these sheet products have shelf lives of several years, so that won't be a factor in the wheel decision here.

Summary

Overall wheel time will generally be driven by a choice between:

- EOQ of the high-volume products
- Shortest wheel cycle time possible based on the available time analysis
- Possibly shelf life issues

The frequency for any specific product will usually be set by:

- EOQ
- Minimum lot size considerations if they apply
- Short-term variability considerations if they apply

We chose the EOQ recommendation, with the 7-day wheel time, for the following reasons:

- It is among the most economical alternatives.
- The weekly regularity is an advantage to operations.
- It provides 24 hours of PIT time each week.
- We can stagger the start of this wheel with the start of other wheels, so that all of the PIT times get evenly distributed over the week.
- It doesn't overload support facilities and resources, as the 2-day wheel did.

Chapter 10

Step 7: Arranging Products—Balancing the Wheel

Now that we have settled on a 7-day wheel, with three products made on every cycle, two made on every second cycle, six made on every third cycle, and two made to order, the next step is to distribute the products across the cycles in such a way that there is a reasonably balanced production demand on each cycle.

The wheel doesn't have to be perfectly balanced; there can be some cycle-to-cycle variation in production requirements. But a reasonably well-balanced wheel offers the following advantages:

- It increases the level of regularity in the manufacturing operation, and levels out the demand for support personnel and facilities.
- The amount of process improvement time (PIT time) available each cycle will be approximately the same, allowing for a uniform window for preventive maintenance and other important tasks.
- It provides a uniform, regularly repeating buffer so that the wheel can be adjusted when the make-to-order products are required.
- It provides a frequently repeating time window that can be used when needed to accommodate unforeseen events and crises, generally resulting in far fewer wheel breakers.

Thus a well-balanced wheel is generally desirable. Occasionally, however, operations may prefer a deliberately unbalanced wheel. If, for example, we have a 7-day wheel with 24 hours of PIT time each week, operations may prefer a 48-hour PIT time block every 2 weeks instead. That would be the case if some of the uses of the PIT time can be much more effectively accomplished in 48-hour blocks than split into 24-hour segments. This will generally not be the case, but it might be, so it's another reason why the perspectives of all stakeholders must be included in product wheel design.

Product	Designation	Weekly Demand D (Rolls)	Frequency (Cycles)	Sequence	Demand per "N" Cycles (Cycle Stock)	Rolls Produced on Each Cycle					
						1	2	3	4	5	6
G	489 J	12	3	1	36			36			36
C	403 J	68	1	2	68	68	68	68	68	68	68
F	406 J	15	2	3	30		30		30		30
L	409 J	9	3	4	27			27			27
A	423 J	130	1	5	130	130	130	130	130	130	130
E	426 J	18	2	6	36		36		36		36
B	403L	108	1	7	108	108	108	108	108	108	108
M	406 L	8	MTO	8	N/A						
D	423 L	26	2	9	52		52		52		52
I	429 L	10	3	10	30			30			30
J	489 R	10	3	11	30			30			30
H	406 R	11	3	12	33			33			33
K	409 R	9	3	13	27			27			27
N	426 R	6	MTO	14	N/A						
	14	440	*Total Rolls per Cycle*			**306**	**424**	**489**	**424**	**306**	**607**
			Total Production Time Required			95.63	132.5	152.8	132.5	95.63	189.7
			Number of Changeovers			3	6	9	6	3	12
			Total Changeover Time			2.25	4.5	6.75	4.5	2.25	9
			Utilization			58%	82.%	95%	82%	58%	118%
			PIT Time			70	31	8	31	70	–31

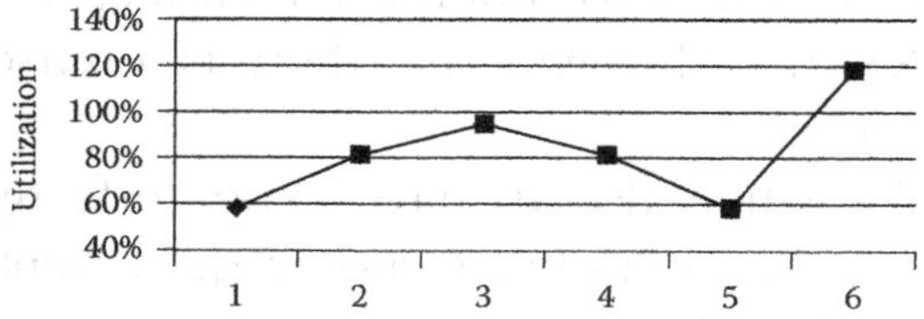

Figure 10.1 Poor balance across cycles.

For the wheel on forming machine 2, the benefits of a balanced wheel outweigh other considerations, so that will be the goal.

If we were to ignore the need for balance, and schedule all second cycle products on cycles 2, 4, 6, 8, and so forth, and all third cycle products on cycles 3, 6, and 9, we would get the result shown in Figure 10.1. This is a highly unbalanced, and in fact unworkable, wheel. The utilization ranges from 58 to 118%; the production requirement on cycle 6 significantly exceeds capacity.

Wheel Resonance

This phenomenon, with all demand coinciding on cycle 6, is an example of what could be called wheel resonance. In scientific terms, resonance occurs when the amplitude of a variable increases significantly when stimulated at its natural frequency. A product wheel might exhibit a somewhat similar phenomenon. Wheels with products made at multiple frequencies will have a natural frequency at the common multiples. For forming 2, a wheel with low-volume products at frequencies of two and three cycles, we see a common multiple at six cycles. At this so-called resonance point all products will be made, thus increasing the production requirement and the utilization for that cycle.

That each cycle can be different with respect to the low-volume products run, and have different capacity utilizations, makes it is necessary to check multiples of the basic cycle frequencies to ensure that there are no production peaks. As another example, if there was one product with a frequency of 5, and it was made on cycles 5, 10, 15, and so on, and a product with a frequency of 6, made on cycles 6, 12, 18, etc., they wouldn't coincide until cycle 30. If the frequency 5 product were made on cycles 4, 9, 14, 19, etc., they would coincide on cycle 24.

Thus it is important to look beyond the first few cycles to ensure that there are no cycles off in the future where an unusually high number of products are due to be made. If the wheel is lightly loaded, this is not much of an issue, but if the wheel is heavily loaded, this can be significant.

If there are several products made at a specific frequency, distributing them so that they don't coincide on specific cycles will help to prevent the problem. For example, with three frequency 5 products, staggering them will avoid the possibility of all of them coinciding with a frequency 6 product at the same time.

With forming 2 products being made on every cycle or every second or third cycles, all possible combinations will be seen within the first six cycles, so any peaks will be obvious from a plot of the first six cycles.

Achieving Better Balance

If we try to find an optimum balance by using our judgment to move products from cycle to cycle by anticipating the effect on overall balance, we might get the alternative shown in Figure 10.2, which is extremely well balanced. Utilizations are uniform, all within an 80 to 84% range. Consequently, PIT time is well balanced, ranging from 27 to 34 hours. So maintenance, product development, and other groups needing PIT time have at least a full 24-hour day each week to schedule their tasks.

Wheels within Wheels

Thus, all other things being equal, the arrangement shown in Figure 10.2 would be a very good choice. However, all other things are not always equal, and a different pattern might result in a better combination of changeovers. When arranging products on the wheel, in addition to finding a good balance, it is critically important to try to group similar products together in a way that makes changeovers easier. In the pattern shown in Figure 10.3, all polymer R products are grouped together on cycles 1 and 4, eliminating the need for an L-to-R polymer change on cycles 2, 3, 5, and 6. (Some users have termed this kind of grouping wheels within wheels.)

Finding a pattern that simplifies changeovers is as important as finding a reasonable balance.

Product	Designation	Weekly Demand D (Rolls)	Frequency (Cycles)	Sequence	Demand per "N" Cycles (Cycle Stock)	Rolls Produced on Each Cycle					
						1	2	3	4	5	6
G	489 J	12	3	1	36			36			36
C	403 J	68	1	2	68	68	68	68	68	68	68
F	406 J	15	2	3	30		30		30		30
L	409 J	9	3	4	27	27			27		
A	423 J	130	1	5	130	130	130	130	130	130	130
E	426 J	18	2	6	36		36		36		36
B	403L	108	1	7	108	108	108	108	108	108	108
M	406 L	8	MTO	8	N/A						
D	423 L	26	2	9	52	52		52		52	
I	429 L	10	3	10	30		30			30	
J	489 R	10	3	11	30	30			30		
H	406 R	11	3	12	33		33			33	
K	409 R	9	3	13	27			27			27
N	426 R	6	MTO	14	N/A						
	14	440			*Total Rolls per Cycle*	**415**	**435**	**421**	**429**	**421**	**435**
					Total Production Time Required	129.7	135.9	131.6	134.1	131.6	135.9
					Number of Changeovers	6	7	6	7	6	7
					Total Changeover Time	4.5	5.25	4.5	5.25	4.5	5.25
					Utilization	80%	84%	81%	83%	81%	84%
					PIT Time	34	27	32	29	32	27

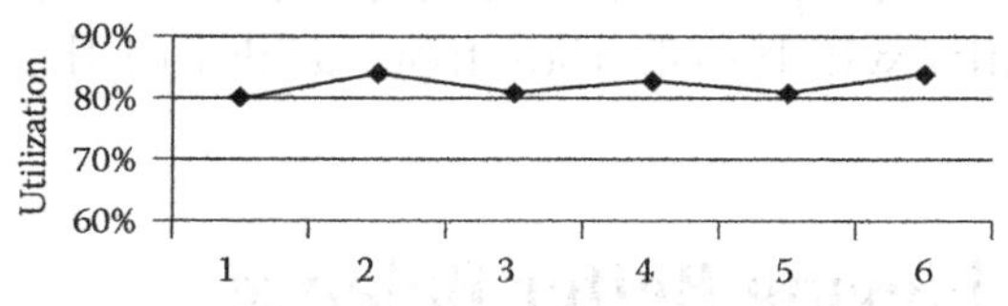

Figure 10.2 A very balanced wheel.

It is very important at this point to get process engineers, operators, mechanics, technicians, and schedulers in a discussion on any other factors that may be relevant in their specific case.

In the case of forming 2, the design team decided that the changeover simplification offered by 10.3 outweighed the need for optimum balance. Thus option 10.3 was chosen, which still gives very good balance, with utilizations ranging from 74 to 90%, and PIT times of at least 23 hours per cycle.

There are several other possible combinations that weren't examined. It is not critical that every possibility be analyzed, only that a reasonable balance that results in a practical combination of changeovers be found.

Summary

In most situations, having the wheel balanced so that the required production is reasonably consistent from cycle to cycle is beneficial. It adds to the regularity that allows operations to develop a repeatable, routine schedule, and provides the benefits of level production described in the early chapters. And it results in a consistent, uniform PIT time for all the essential tasks that the PIT time can be used to accomplish.

Product	**Designation**	**Weekly Demand D (Rolls)**	**Frequency (Cycles)**	**Sequence**	**Demand per "N" Cycles (Cycle Stock)**	**Rolls Produced on Each Cycle**					
						1	2	3	4	5	6
G	489 J	12	3	1	36			36			36
C	403 J	68	1	2	68	68	68	68	68	68	68
F	406 J	15	2	3	30		30		30		30
L	409 J	9	3	4	27			27			27
A	423 J	130	1	5	130	130	130	130	130	130	130
E	426 J	18	2	6	36		36		36		36
B	403L	108	1	7	108	108	108	108	108	108	108
M	406 L	8	MTO	8	N/A						
D	423 L	26	2	9	52	52		52		52	
I	429 L	10	3	10	30		30			30	
J	489 R	10	3	11	30	30			30		
H	406 R	11	3	12	33	33			33		
K	409 R	9	3	13	27	27			27		
N	426 R	6	MTO	14	N/A						
	14	440	*Total Rolls per Cycle*			**448**	**402**	**421**	**462**	**388**	**435**
			Total Production Time Required			140	125.6	131.6	144.4	121.3	135.9
			Number of Changeovers			7	6	6	8	5	7
			Total Changeover Time			5.25	4.5	4.5	6	3.75	5.25
			Utilization			86%	77%	81%	90%	74%	84%
			PIT Time			23	38	32	18	43	27

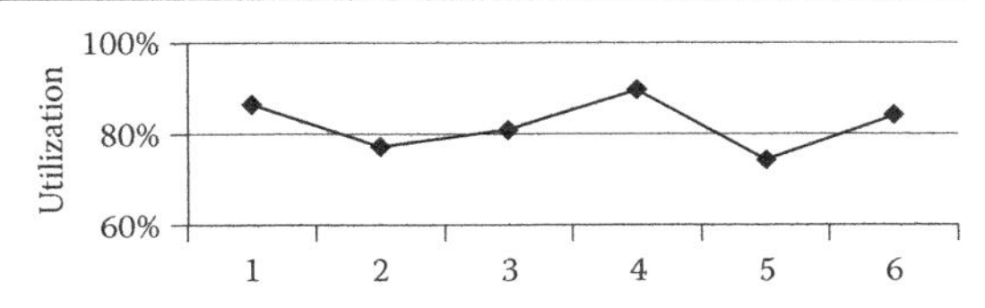

Figure 10.3 Like products grouped by cycle.

It is not necessary to find a perfectly balanced arrangement of products; close enough is good enough. And it is important to find a pattern where similar products are made on the same cycle to reduce changeover time or cost, like the case shown in Figure 10.3. There are sometimes other operational factors and influences that must also be considered. Selecting the best arrangement alternative can come down to a judgment call, and like all judgment calls in product wheel design, the more perspectives and experiences that are included in the discussion, the better the judgment will be.

Chapter 11

Step 8: Plotting the Wheel Cycles

Once an initial product wheel design has been generated, it must be reviewed with all whose jobs will be affected by it, and all who must approve any aspect of it. (This will be discussed in more detail in Chapter 13.) All of them must understand the wheel concept and how it will change current work processes and systems.

In order to create that understanding, a visual representation of the wheel will be extremely helpful, offering these benefits:

- It provides an excellent tool for communicating the wheel concept, the attributes of wheels, and the specific details of how it applies to this manufacturing process.
- It gives people a clear image of the specific schedule or schedules to be followed in their operation.
- It serves as a thought provoker for discussion and questions.

It also adds to the understanding of the people who have been actively involved in design of this wheel. A diagram creates a much clearer impression of how the wheel will actually operate than the numbers on the spreadsheets they have been working with.

If a wheel has low-demand products at several frequencies, such that there are a number of unique cycles, it is not necessary, or even useful, to plot the wheel for every one. A few representative cycles are usually enough to explain the concept, to clarify that cycles can be different, and to generate discussion and questions.

The wheels for forming 2 on cycles 1, 2, 4, and 5 are shown in Figure 11.1, which highlights the differences between cycles. If these were in color, the visual effect would be much more dramatic.

If the wheels were developed using spreadsheets in an application like Microsoft Excel, plotting the cycle pie charts will be easy. Just remember to add a row to the spreadsheet for the process improvement time (PIT time).

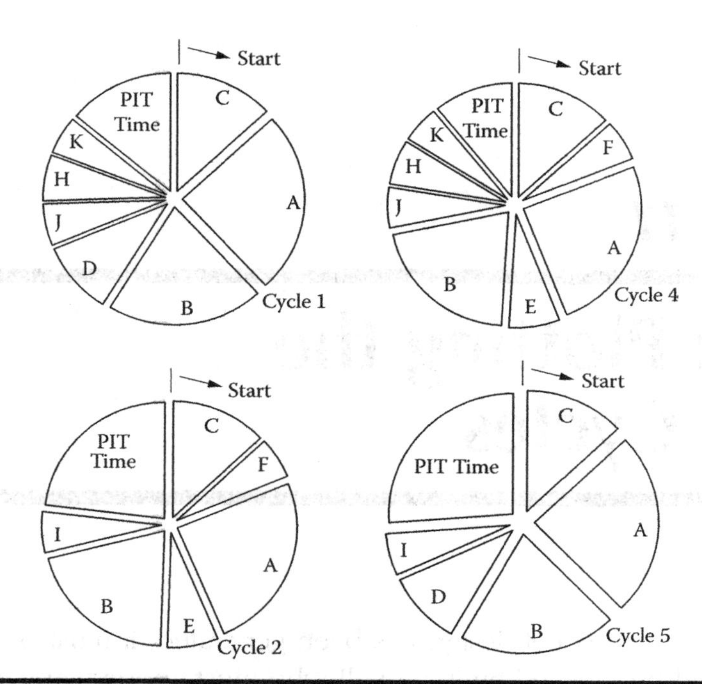

Figure 11.1 Forming 2—Various wheel cycles.

Wed	Thur	Fri	Sat	Sun	Mon	Tue	Wed

C	A	B	D	J	H	K	PIT Time	C

|← Cycle 1 → |→ Cycle 2 →

Figure 11.2 Bar chart for wheel cycle 1.

I've also seen linear bar charts used, and they do tell the story visually, but more people seem to grasp the idea when presented with a circular pie chart. The linear chart does have the advantage that it is easier to overlay a timeline, as shown in Figure 11.2. The choice of a specific visual format is up to the product wheel team, based on their ideas on what will convey the concept and schedules to their audience most clearly.

More people seem to prefer the circular representation, and some plants have adopted the circular product wheel symbol as a logo they use on all formal communications with the manufacturing force about product wheels.

Summary

It is often said that a picture is worth a thousand words, and that is certainly true of product wheels. A graphic representation of the wheel builds greater understanding, and therefore a higher level of commitment to the concept and to implementation. The specific format and style used is not critical, but developing some type of visual image is.

Chapter 12

Step 9: Calculate Inventory Requirements

Some inventory is generally needed to support a product wheel. For the make-to-stock products, the inventory required will be roughly proportional to wheel time. For these products, there must be inventory downstream, either as work in process (WIP; aka semifinished inventory) or as finished product, to satisfy needs for each material during the period when other products are being produced on the wheel. Figure 12.1 tracks the inventory for a single product made on a wheel, as a function of time. An amount, P1, of that product is produced at the beginning of wheel cycle 1, and flows through downstream process steps into the inventory represented by the figure. Demand for that product during the remainder of the wheel cycle will consume portions of that inventory until the next production, P2, arrives at the inventory. This rising and falling of inventory in a sawtooth pattern repeats for every cycle of the wheel, but will have some variation depending on the demand during any cycle.

With make-to-order products, inventory may or may not be needed. As product orders are received, they are loaded onto the wheel schedule. After they are produced on the wheel, they flow through the downstream steps, through to packaging, truck loading, and shipment to the customer. They may stop temporarily at a buffer inventory, but for reasons other than requirements imposed by the wheel schedule. The exception is for orders where the customer has requested a specific delivery date. Even though the customer lead time is long enough to allow us to wait until the appropriate time comes around on the wheel to produce the material, once it is produced it may have to wait in inventory until the requested ship date.

It is important to know how much inventory will be needed to support the wheel, so all stakeholders can agree that this is a good use of working capital. If any of the business leadership has a concern about the amount of inventory

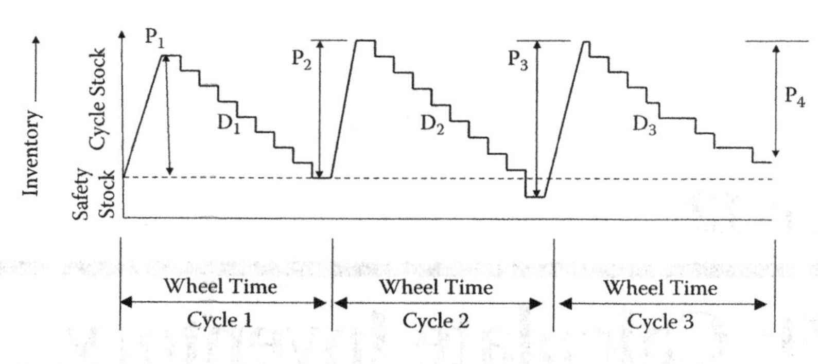

Figure 12.1 Inventory required to support the wheel—Cycle stock and safety stock.

required, they should be given alternatives and the economic consequences of each. The inventory will generally be less than the current inventory, but may be more than some people were expecting. Inventory requirements can often be lowered by shortening the wheel cycle, but at the cost of more changeovers. The calculations we've gone through so far provide the basis for calculating the economic ramifications of any hypothesis.

The inventory requirement for each product is also needed so the current level can be adjusted to the appropriate level prior to starting the wheel.

Once the required inventory has been calculated, it should be shown on the future state value stream map in the appropriate inventory data box, so it is clear what the new target is.

Inventory Components

The required inventory can be considered to be of two components: cycle stock and safety stock. Cycle stock is the inventory needed to support downstream demand while all other products are being made. In other words, it is the stock needed to satisfy demand during the production of that material and from the completion of the spoke until that spoke comes around again. So for a product made every wheel cycle, it is the average demand for the product during the wheel time. For products made less frequently, it is the average demand over a wheel cycle times the frequency factor.

The equation for calculating cycle stock is

Cycle stock = Average demand per unit time × Wheel time × Cycle frequency.

Looking at how this applies to products made on forming machine 2, product G has a weekly demand of 12 rolls and a frequency of 3 on the 1-week wheel. Thus its cycle stock is 36 rolls.

Figure 12.2 illustrates the fact that shorter wheels require less cycle stock, and longer wheel times more cycle stock. If the average demand is 1000 pounds per

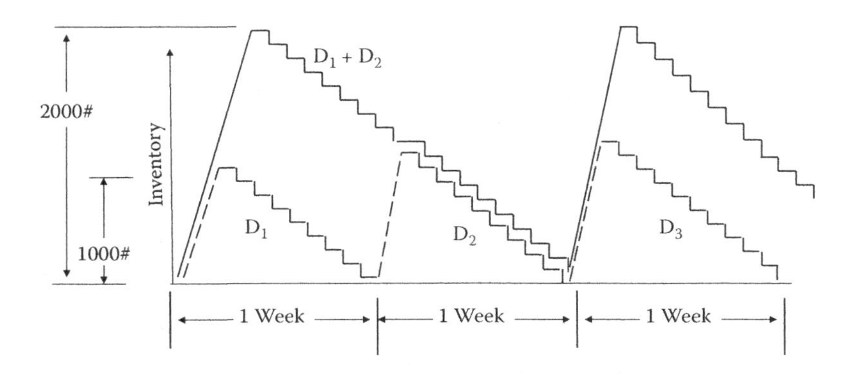

Figure 12.2 Inventory needed increases with longer wheel time.

week, and we make the product weekly, the cycle stock is 1000 pounds. If we produce every 2 weeks, the cycle stock becomes 2000 pounds. A more thorough discussion of cycle stock can be found in Appendix A.

If demand were always exactly equal to the average, only cycle stock would be needed. However, demand will typically vary in a random fashion, sometimes greater than average and sometimes less. To avoid stock-outs during periods of higher demand, safety stock is usually carried. The demand, D2, in period 2 of Figure 12.1 is higher than average, and the safety stock provides material to supply downstream needs and customers until P3 is produced on the next cycle. Safety stock can also provide protection from stock-out in cases where the wheel, due to production problems, doesn't return to this spoke at the scheduled time.

The amount of safety stock required is a function of the variability in demand we are trying to cover, and the desired customer service level. Ninety-five percent is a typical service level goal, meaning that we want a stock-out of this specific product on no more than 5% of the wheel cycles. Of course, we can set higher service level targets, but they will require more safety stock, so 95% is often viewed as a reasonable compromise.

If variability in demand can be expressed as a standard deviation (σ_D), safety stock can be calculated by

$$\text{Safety Stock} = Z \times \sigma_D$$

where Z is the factor representing the service level goal. For a 95% service level target, the Z factor is 1.65, so if the target for product A on forming machine 2 is 95%, the safety stock requirement will be 46 rolls (1.65 × 28 rolls). A more thorough discussion of safety stock can be found in Appendix B.

Total Inventory Requirements

Using these equations, the total inventory required to support the product wheel can be calculated. This includes the cycle stock and safety stock for all make-to-stock products made on the wheel. Figure 12.3 tabulates total inventory for forming

Product	Designation	Weekly Demand D (rolls)	σ_D (rolls)	Frequency (cycles)	Demand per N Cycles (cycle stock)	CSL Target (%)	σ_D (rolls/N cycles)	Safety Stock (rolls)	Peak Inventory (rolls)	Average Inventory (rolls)
G	489 J	12	4	3	36	95%	6.9	11	47	29
C	403 J	68	17	1	68	95%	17	28	87	58
F	406 J	15	3	2	30	95%	4.2	7	36	22
L	409 J	9	3	3	27	95%	5.2	9	35	22
A	423 J	130	28	1	130	95%	28	46	145	96
E	426 J	18	3	2	36	95%	4.2	7	42	24
B	403 L	108	18	1	108	95%	18	30	116	73
M	406 L	8	6	MTO	N/A	N/A	N/A	N/A	N/A	0
D	423 L	26	5	2	52	95%	7.1	12	61	36
1	429 L	10	2	3	30	95%	3.5	6	35	20
J	489 R	10	2.7	3	30	95%	4.7	8	37	22
H	406 R	11	3	3	33	95%	5.2	9	41	25
K	409 R	9	3.5	3	27	95%	6.1	10	37	23
N	426 R	6	4	MTO	N/A	N/A	N/A	N/A	N/A	0
									Total	450

Figure 12.3 Forming machine 2—Total inventory required to support the wheel.

machine 2. It should be noted that the total inventory at the completion of a production spoke for a product may not quite equal cycle stock plus safety stock. This is because some of the material produced was consumed during its spoke. This can be significant for products occupying a large portion of the wheel: product A occupies 24% of a full wheel cycle, so approximately 24% of its production is consumed before the end of the A spoke. Thus while its cycle stock is 130 rolls, 31 of those get consumed during the production, so the inventory at the end of that spoke will be 99, plus the safety stock of 46 rolls, for a total of 145 rolls. Its average inventory will be one-half of the 99 rolls, plus the safety stock, for a total of 96 rolls:

$$\text{Peak Inventory} = \text{Cycle Stock} \times \left(1 - \frac{\text{Demand}}{\text{Production Rate}}\right) + \text{Safety stock}$$

$$\text{Average Inventory} = \frac{1}{2} \times \text{Cycle Stock} \times \left(1 - \frac{\text{Demand}}{\text{Production Rate}}\right) + \text{Safety stock}$$

Forming 2 has a weekly production rate of 540 rolls, so the specific equations for product A are:

Peak inventory = 130 rolls × (1 – 130 rolls/540 rolls) + 46 rolls = 145 rolls

Average inventory = ½ × 130 rolls × (1 – 130 rolls/540 rolls) + 46 rolls = 96 rolls

Figure 12.3 shows the total inventory required to support forming machine 2's wheel to be 450 rolls. The total inventory will always be at a level approximately equal to the sum of the averages for each product, because there will always be some products near the peak and others near the low point on their individual sawtooth waves.

Inventory Benefit of the Wheel

The 450-roll inventory requirement represents a 10% reduction from the prior need for 500 rolls. While that in itself is significant, a working capital reduction of approximately $100,000, the greater benefit may be in the stability it brings to the operation. With a wheel plan that is rarely broken, we have far fewer unplanned changeovers and less scrambling to deal with the schedule disruptions that had occurred frequently in the past. The inventory and working capital reductions are very attractive, but over time, people tend to appreciate and value the predictability and dependability more.

Prior to the wheel, we had applied the virtual cells shown in Figures 5.1 and 5.2, which also allowed a significant inventory reduction. The assignment of

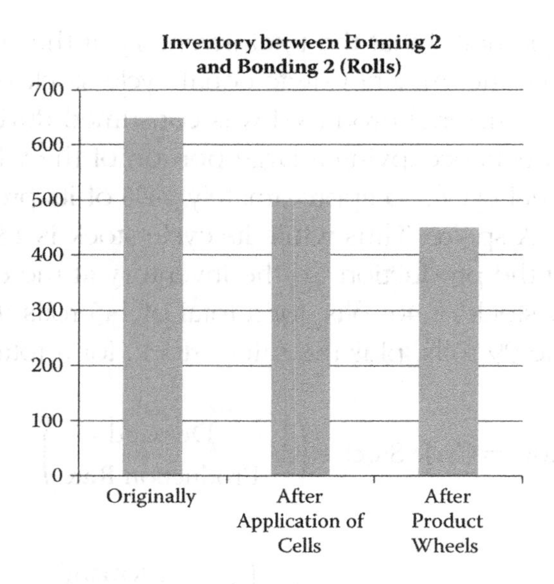

Figure 12.4 Inventory reduction with cellular flow and product wheels.

each product to a specific cell, called Group Technology, and the determination of inventory requirements based on mathematical principles rather than gut feel brought this inventory down from 650 rolls to the 500-roll level. Figure 12.4 illustrates the two-step inventory reduction.

Seasonality

In calculating the safety stock required, it is important to understand whether there is any seasonality causing some of the variation in demand. The demand for sun tan lotion, for example, has a very high seasonal component, rising to very high levels in the summer months and declining during the colder months. If there is a seasonal component in the demand variation, recognizing it and filtering it out can reduce the safety stock requirements significantly. Seasonality refers to any predictable variation in demand, which may repeat on an annual cycle, or something less, perhaps monthly or even weekly. The breakfast cereal example described in Chapter 8 is an example of shorter-term seasonality, with its pattern repeating over a 4-week period.

To understand the method commonly used to filter seasonality out of a demand pattern, we will use a tomato ketchup bottling plant as our example. Ketchup sales see a variation in consumption over the course of any year. Sales peak in the summer months when people are enjoying picnics and cookouts, and decline somewhat in the months when people tend to stay indoors. The plant where this is produced and packed has eight lines, which make ketchup, barbecue sauce, and other condiments packed in bottles of various sizes, ranging from very small bottles for room service use at hotels to multigallon containers for institutions and cafeterias.

Month	*2007* Demand	*2008* Demand	*2009* Demand	*2010* Demand
Jan	126	145	125	138
Feb	144	112	127	134
Mar	143	114	135	133
Apr	130	125	122	122
May	188	175	169	159
June	185	158	174	180
July	167	182	190	184
Aug	167	165	151	165
Sept	135	139	147	132
Oct	134	117	143	122
Nov	150	115	149	137
Dec	130	141	113	145

Figure 12.5 Monthly demand for tomato ketchup in 24-ounce bottles (thousands of cases).

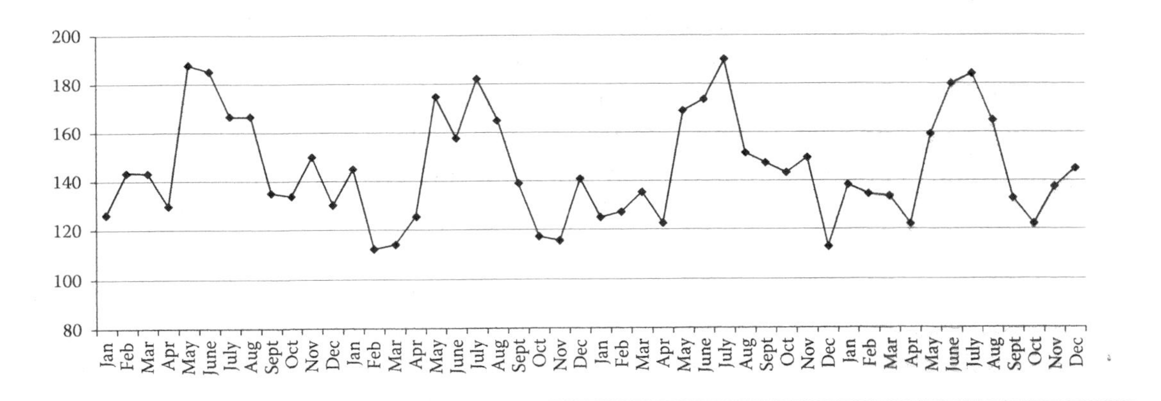

Figure 12.6 Forty-eight-month demand profile—Tomato ketchup in 24-ounce bottles (thousands of cases).

To focus on one specific product, 24-ounce bottles of ketchup packed in cases of 12 bottles each occupy approximately 15% of the capacity of one of the high-speed lines. Figure 12.5 shows 4 years of demand data recorded on a monthly basis. The graph in Figure 12.6 illustrates this and shows the peak in each summer season. May through August have much higher than average demand, while the other eight months see below-average demand.

This variation has a standard deviation of 22,000 cases, so if we are to cover all of it with safety stock, we would need 36,300 cases of safety stock (1.65 × 22,000). If the seasonality is filtered out, far less safety stock will be needed.

Figure 12.7 shows the results of filtering out the seasonal component of demand variability following these steps:

1. Average the demand for all 48 months, resulting in 146,000 cases per month.
2. Average the four January months, equal to 134,000 cases.
3. Divide the January average by the 48-month average, giving a factor of 0.92. This means that, on average, demand in January is 8% lower than average.
4. Divide each January demand value by the 0.92 factor to get a seasonally adjusted value for each year's January demand. This raises each January data point by 8% to indicate what it would be without the seasonal variation.
5. Repeat for each of the remaining months. This will show that July, at 181,000 cases, is 24% higher than average, so each July data point is divided by 1.24 to reduce it to the level without the seasonal variation.

Figure 12.8 plots the 48-month demand history with the seasonal variation filtered out, and the original data shown as a dotted line for comparison. The random variation is still there, but the seasonal variation is not.

Having done this analysis, we will set different cycle stock levels for the peak season and the low-demand season. Cycle stock will be set at 172,500 cases for each of May, June, July, and August, and at 132,000 cases for the remaining months. These represent the actual demand averages for each of the two seasons.

The filtered data show a standard deviation of 10,000 cases, versus 22,000 cases for the unfiltered data, so the safety stock will be less than half what would have been required had we not recognized the seasonal trend.

It must be emphasized that the purpose in calculating seasonally adjusted demand values is only so that we can calculate the true random variation and set the safety stock appropriately. Cycle stock will be set for each season by the actual average demand for that season.

If seasonality is present, and treated this way, economic order quantity (EOQ) should be recalculated for each season. Because of the flatness of the EOQ curve described in Chapter 8, moderate amounts of seasonality won't change the EOQ recommendation, but for extremely seasonal items like lawn mowers or suntan lotion, it likely will. It wouldn't be unusual for EOQ to suggest a 1-week frequency during the peak season, and a 2- or 4-week frequency during the off-season.

Customer Lead Time

Everything we've done so far assumes that we must have enough inventory in stock to meet any reasonable demand instantaneously. Our premise is that we have enough finished product inventory so that orders are shipped as they are received. In some cases, however, customer lead time expectations are long enough that we can take advantage of a portion of that lead time to schedule the wheel in a way that reduces our inventory requirement.

Month	2007 Demand	2008 Demand	2009 Demand	2010 Demand	Monthly Average	Seasonality Factor	Seasonally Adjusted Demand			
							2007	2008	2009	2010
Jan	126	145	125	138	134	0.92	138	158	136	151
Feb	144	112	127	134	129	0.89	152	126	143	151
Mar	143	114	135	133	131	0.90	158	126	150	148
Apr	130	125	122	122	125	0.86	151	146	143	142
May	138	175	162	153	173	1.19	158	147	142	134
June	185	158	174	180	174	1.20	155	132	145	151
July	167	182	130	184	181	1.24	134	147	153	148
Aug	167	165	151	165	162	1.11	150	148	135	148
Sept	135	139	147	132	138	0.95	142	146	155	139
Oct	134	117	143	122	129	0.89	151	132	161	138
Nov	150	115	149	137	138	0.95	158	122	158	145
Dec	130	141	113	145	132	0.91	143	155	124	160
			4-year average		146			4-year average		146
			4-year standard deviation		22			4-year standard deviation		10

Figure 12.7 Monthly demand with seasonality filtered out (thousands of cases).

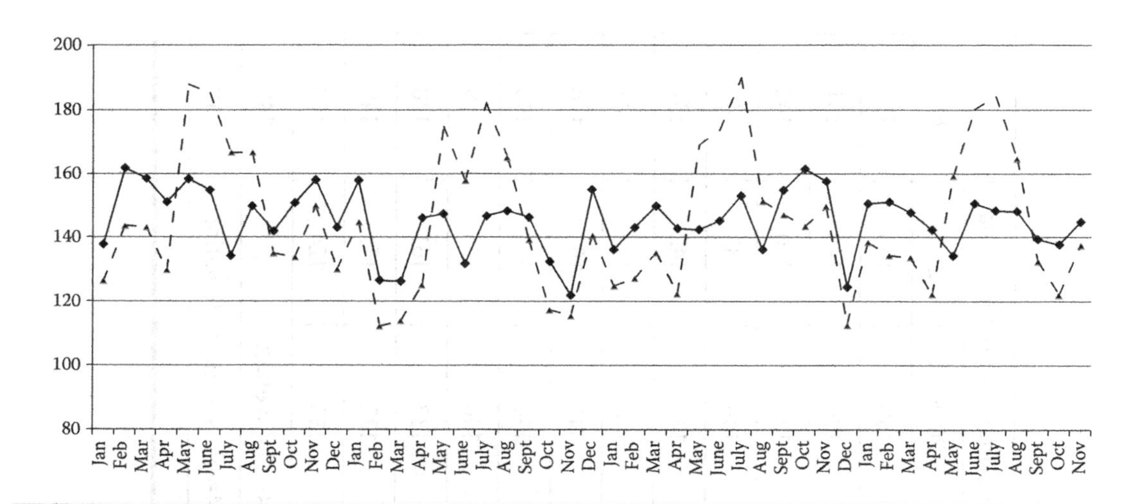

Figure 12.8 Forty-eight-month demand profile with seasonality filtered out (thousands of cases).

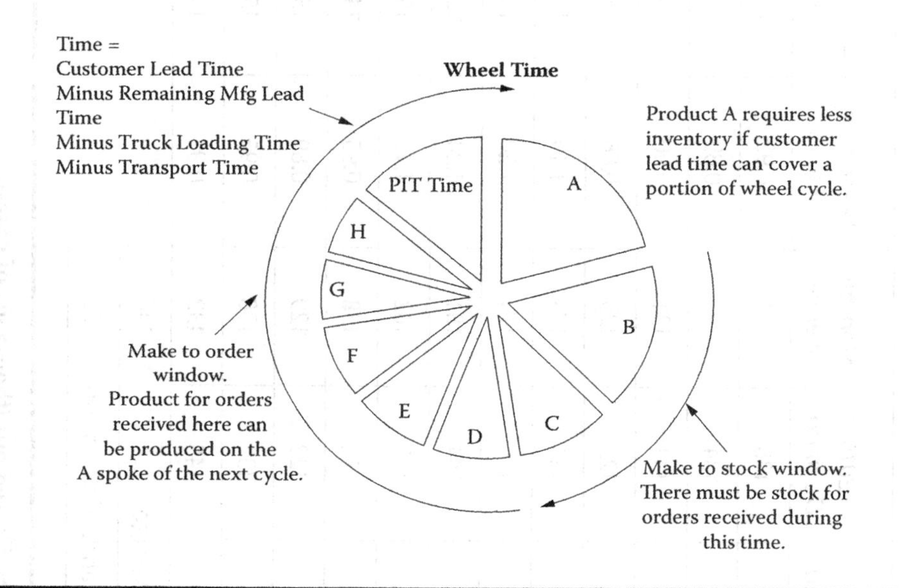

Figure 12.9 Customer lead time reduces time requiring inventory.

For example, if customers expect product within 10 days, and it takes us 2 days to complete the manufacturing steps between the wheel step and shipping, 1 day to load and dispatch the truck, and 3 days to arrive at the customer, we have 4 days (10 – 2 – 1 – 3 = 4) to manufacture the material if it is not in stock when the order is received.

That allows us to make any orders that come in 4 days or less before the start of a spoke to order rather than carrying inventory. The concept is illustrated in Figure 12.9. Considering product A, for example, and that the above lead time parameters apply, and that we have a 7-day wheel where spoke A occupies approximately 1 day, we must end the A spoke with enough inventory to satisfy the demand for the next 2 days. Any orders that come in after that are within 4 days of the next A spoke, so they can be filled on the next cycle. So we are making stock for the orders expected during the first 3 days of the cycle, and holding the last 4 days of orders to be made to order on the next cycle. Thus while we're in spoke A, we are making the last 4 days of orders on a make-to-order basis and the anticipated orders for the next 3 days on a make-to-stock basis.

Admittedly this complicates wheel scheduling somewhat, so it is generally not worth considering unless the excess lead time is significant. If it amounts to several days, it may give a sizable inventory reduction. But you must remember to include the lead times for all of the downstream steps beyond the wheel step in the subtraction from customer lead time commitment.

Summary

Determining inventory requirements is an important step in product wheel design. It allows the business leadership to understand how much working capital will be required, and engage in a discussion of the economic consequences of any variations on the suggested wheel design. After agreement is reached, it provides targets for the inventory ramp-down or ramp-up required before the wheel can begin. (The current inventory levels will generally be higher than the required levels, but sometimes there are a few products at inappropriately low levels.)

The inventory for each product will consist of cycle stock and safety stock. Cycle stock is the inventory required to carry a product from spoke to spoke. Safety stock is to protect against variability in demand, specifically those cycles where demand is higher than average. The amount required depends on the degree of variability present and on the desired service level.

If the variability includes seasonality, it should be filtered out so that only the safety stock needed to cover true random variability will be carried. The seasonal portion of the variation can be covered by setting the cycle stock to the average demand for the current season.

If customers allow delivery times significantly in excess of the manufacturing lead time after the wheel step plus the loading and transport time, it is worth considering taking advantage of that time by employing a partial make-to-order strategy and thus lowering inventory requirements.

Chapter 13

Step 10: Review with Stakeholders

Once an initial product wheel design has been generated, it should be reviewed with all stakeholders, which includes all those whose performance is impacted by how well the wheel functions, representatives of all groups who will interact with wheel execution in any way, and all who must approve any aspect of it. Their input and perspectives may uncover concerns about the design, which must be addressed. Product wheel design is often an iterative process, and so you should expect that you may need to loop back to earlier steps. The good news is that the techniques described in the earlier chapters make adjustments and revisions to the design relatively straightforward.

What to Review

The topics to be covered in any review depend on the knowledge base of the audience members and how closely they've stayed in touch with the wheel design as it progressed. Possible topics include:

- The product wheel concept
- The product sequence chosen
- The wheel time and the key factors influencing it
- The distribution of lower-volume products across alternate cycles
- Products to be changed to a make-to-order (MTO) strategy
- Inventory requirements
- The process improvement time (PIT time) concept and its value
- The implementation plan and target turn-on date

The wheel graphics described in Chapter 11 will be helpful in these reviews, enabling people to better visualize wheel operation. Because wheel length and inventory requirements are likely discussion topics, charts should be prepared showing the economic order quantity (EOQ) results and perhaps several alternate scenarios. A list of PIT time requirements and uses will reinforce the value of this time and the importance of using it wisely.

Who to Include

People who are directly involved in executing the wheel:

- Production planners and schedulers: These are the people who generate specific schedules in accordance with the wheel design on an ongoing basis.
- Operators: They of course operate the wheel, and enter performance to plan data and reasons for nonconformance on the display boards.
- Raw material coordinators: Their planning process might require modification to accommodate the schedule differences caused by the introduction of the wheel design.

People whose work is affected by the wheel schedule:

- Maintenance supervisors, mechanics, and electricians: If they are involved in changeovers, they must be informed that the frequency of changeovers may be increasing or decreasing.
- Quality lab supervisors and technicians: If there is any testing required after a changeover, either to ensure that the equipment is free of contaminants or that quality is again within specification, the laboratory personnel must be informed that the frequency of changeovers may be increasing or decreasing.
- Someone from the IT department: It is very likely that changes will have to be made to the plant enterprise resource planning (ERP) or manufacturing resource planning (MRP II) computer systems, or whatever type of system or process is currently used to plan and schedule production. Most major ERP systems, like SAP and Oracle, can accommodate product wheel scheduling, but it is not always obvious or straightforward how to configure the system for wheels, so the plant IT group may have a significant role to play in product wheel implementation.
- New product development engineers: If they need line time for product development tests and qualification runs, they should know that there will be some regularly recurring time block that may be available for their use, and understand the process for scheduling it.
- The plant project group: Similarly, if the project team needs time to install new equipment or to modify existing equipment, they must be informed that time can be made available for that, and how to arrange for it.

- Warehouse supervisors: Must have an understanding of how inventory requirements will change, so they can decide how to accommodate it. For example, if warehouse space is being leased on a square footage basis or on a rack occupancy basis, they must coordinate any decreases or increases with the warehouse contractor.

People who must approve any aspect of the wheel design—any changes to the current scheduling methodology could change things in which they have an interest:

- Business director: May have concerns about inventory carrying cost and about customer delivery performance. Approval from someone at this level will likely be required if any products are suggested for MTO.
- Marketing manager: Similar to the business director, concerns about customer delivery performance and moving products to MTO.
- Operations director: May have concerns about inventory carrying cost, changeover cost, resources required for changeovers (operators, mechanics, lab facilities), yield losses after restart, and a number of other things.
- Plant manager: Will likely share whatever concerns the operations director has.
- Technology manager: May be responsible for yield metrics, and therefore be concerned about yield losses on restart, in cases where the number of changeovers will increase.

Possible Concerns and Challenges

The possible challenges to the wheel design include:

- The wheel is too short: There are too many changeovers. This may simply be a knee-jerk reaction from operations, not wanting to have more changeovers to do. But you've done the math and can show that this wheel time is feasible and economically beneficial.

  This challenge may come from maintenance, based on a very legitimate concern about the lack of enough mechanics to perform all the changeovers. If that concern is raised, you should calculate the cost of adding a mechanic or two, versus the cost of lengthening the wheel.

  It may come from the quality lab, concerned about the increased product testing that normally comes from increasing the number of changeovers. Again, the economics of adding lab technicians or instruments should be weighed against the cost of inventory if the wheel is lengthened.

  Because there are often yield losses on restart, the quality manager may be concerned about the reduction in yield metrics accompanying a higher number of changeovers. But if EOQ principles were used to set wheel time, yield losses have already been taken into account, and the current wheel time is the optimum balance between yield losses and inventory costs.

- If the wheel is very short, there may not be enough PIT time for all of the new product qualification runs that the technical group would like to do, for equipment additions and modifications that the project group has planned, and for all of the other needs, including required PMs (preventive maintenance) and operator training. If there are more legitimate needs than there is PIT time available to accomplish, the wheel should be lengthened to provide adequate time for these important tasks. But the PIT time requirements should be evaluated to ensure that they are necessary, enough to warrant lengthening the wheel. And ways to accomplish these tasks without consuming line time should be explored.
- The inventory is too high. This is not generally a concern, because the total inventory requirement usually goes down with product wheels, but if inventory must be increased, it may raise questions. They may come from the business manager, who doesn't want to carry that much inventory cost. They may come from the warehouse manager, who is concerned about the storage space required.

  If the wheel design requires inventory to increase, be prepared to explain the factors that led to that recommendation. Charts showing the EOQ results, trade-offs between inventory cost and changeover cost, and the ramifications of shortening the wheel will be very helpful.

  If the decision is that inventory must be reduced, shortening the wheel is generally the most practical way. But if you were already close to the fastest wheel possible, this may not offer much relief. If some of the lower-volume, high-variability make-to-stock products can be made to order, that will help reduce inventory.

  The worst thing to do is to keep the wheel design and all parameters where they are, and simply trim the inventory. That will almost certainly lead to stock-outs, confusion, chaos, and the loss of the wheel structure and discipline.
- If there are low-volume, high-variability products being suggested for a make-to-order strategy, the marketing manager may have a concern about customers being willing to accept longer order fulfillment lead times for those products. But this should have been reviewed with the marketing organization and the business director and resolved in step 3.

Many of these concerns can be alleviated by either having representatives of all affected groups on the design team or having frequent progress reviews to try to get the issues on the table as quickly as possible. If your plant has a culture of making improvements using kaizen events, product wheel design makes an excellent scope for an event. A well-designed event will ensure that most affected groups or functions are represented in the design process. (Kaizen events were briefly covered in Chapter 3.)

Summary

The simple fact that the scheduling methodology is being completely altered may have some effect, either directly or indirectly, on most people in the operation. Therefore they should clearly understand the changes and how they will be affected. This includes the people who interact with wheel operation as part of their normal responsibilities. It also includes people in management who must be in alignment that any increases in cost, such as changeover costs, are justified by the benefits of wheels, like reduced inventory, faster response, reduced schedule disruption, and improved customer delivery performance.

But don't wait for a final review to get any changes in front of the appropriate stakeholders. If their perspectives and priorities are incorporated early in the design, it will likely avoid most of the rework.

Chapter 14

Step 11: Assign Responsibility for Allocating PIT Time

We've spoken a number of times about the importance of process improvement time, commonly called PIT time, and all of the valuable uses for any time on the wheel that is not needed for production or changeovers. In order to ensure that the time is used to best advantage, and that none of it is wasted, it is important to give someone the responsibility for understanding all of the appropriate uses, to allocate the time appropriately, and to lead the reconciliation of any conflicts.

To repeat a point made in Chapter 2, referring to this time as process improvement time rather than slack time creates a much more proactive mindset about the value of this time and the need to make the best use of it. Calling it slack time leads to a very casual attitude about putting any effort into deciding how it should be used.

I realize that the term *PIT time*, process improvement time time, is redundant, but teams seem to think that *PIT time* is a more appropriate description than *PI time*.

The responsibility for understanding all needs and allocating the time accordingly can be given to a single individual or to a small team. This responsibility includes documenting and communicating the planned use of each block of PIT time.

Appropriate Uses of PIT Time

There are a number of important tasks requiring line time that can be done during the PIT time. This list is certainly not complete, for there can be unique needs in specific processes, nor is it in any priority.

- Routine preventative maintenance (PM) tasks
- Equipment modifications and upgrades

- Installation and checkout of a new piece of equipment or control system element
- New product development and qualification runs
- 5S activities, to improve workplace organization, visibility, and housekeeping
- SMED activities to simplify and shorten changeovers
- Operator training
- Kaizen events aimed at flow improvement, bottleneck reduction, reliability improvement, quality improvement, and defect reduction

If a wheel is relatively lightly loaded, with a large spoke of PIT time on each cycle, it is not necessary to completely fill all of the available time. But if time is left idle, it should be as the result of a coordinated analysis of needs, not from lack of attention.

Chapter 15

Step 12: Revise the Scheduling Process

Before wheel operation can begin, all formal scheduling processes must be examined and modified as appropriate to accommodate product wheel scheduling. This includes the scheduling processes executed within plant and corporate computer systems, and any visual scheduling tools being used in the production area.

Wheel Concepts and the Production Scheduling System

Once the wheel design has been finalized and approved, any computer systems used for production planning and scheduling must be reconfigured to accommodate wheel scheduling. This may include an enterprise resource planning (ERP) system, such as SAP, Microsoft Dynamics, Infor, NetSuite, or JD Edwards, or one of the other systems available from Oracle, used at the plant level or at the corporate level. Most of these systems can be set up for product wheel methods, but the ERP training material can make this challenging. These systems tend to be designed around a material requirements planning (MRP) strategy, where production tends to be scheduled by due dates for end items and lead times and bill of material breakdowns for dependent demand items (component parts or materials for the end item). The documentation and training for most ERP systems emphasize an MRP mode of operation.

Product wheels, which schedule production in lots grouped together within a fixed campaign cycle and based on a particular sequence, rather than on due dates for individual orders, can be implemented in most ERP systems, but it takes someone with a good understanding of the system to configure it for this strategy. Therefore it is usually beneficial to include someone from the IT function in the wheel design to some degree. He or she doesn't necessarily have to be a full-time member of the team, but he or she must understand the underlying philosophy of product wheels, and the details of the current design.

In some cases, an ERP system is used for higher-level accounting and business processes, and for demand forecasting and sales tracking, but the details of scheduling the equipment to meet production goals are done using self-developed spreadsheets in Microsoft Excel. Altering them to incorporate wheel strategy and parameters is generally much more straightforward, and should be done by the developer of the spreadsheet or its current user.

Visual Management of the Current Wheel Schedule

A key component of product wheel scheduling is a visual display of the current schedule, located in the production area and updated in real time, so that everyone involved can easily tell where you are on the wheel, what product is next, and whether you are ahead of or behind schedule.

If visual scheduling tools are not currently being used on the plant floor, they should be added, both as a valuable tool in product wheel management and as an important step forward in the workforce engagement process. An example of a visual scheduling board for forming machine 2 is shown in Figure 15.1. Such a display is sometimes called a takt board, because it displays the production required to meet the current level takt rate, and also provides operators a place to record performance to the takt rate.

The board should show the next cycle of the product wheel, or at least a few days of it, the products to be made on this cycle, and the quantity of each. It should have an area where operators can record performance to the plan as the wheel progresses, so that all in the area know the current status and so that corrective action can be taken if problems are occurring.

In the example, production for several days is shown in 4-hour blocks. The specific product type to be made and the takt quantity for that time block are listed. Actual production, and whether it was greater than, equal to, or less than takt, is plotted as a cumulative number. Reasons for producing more or less than the takt quantity are listed, so that corrective action can be taken for chronic problems. This chart shows that a changeover to product 423 J was completed on Monday morning at 12:00 a.m., that it went well, and that the machine was running well, so production exceeded takt for the first two time periods. Therefore an adjustment was made, and production was deliberately lower than takt for the next two periods to avoid overproduction on this campaign. A changeover to 403 L was done at 4:00 p.m. and didn't go very well, so production fell behind takt. Because of that, and a sheet tear-out during the midnight to 4:00 a.m. time period, extra rolls had to be produced during the later portion of this campaign to catch up to the takt requirement.

A board like this:

- Allows all of the operators to know what product is to be produced
- Captures actual production quantity and how it compares to the schedule

Day	Monday						Tuesday						Wednesday	
Time	12–4 AM	4–8 AM	8 AM–12	12–4 PM	4–8 PM	8 PM–12	12–4 AM	4–8 AM	8 AM–12	12–4 PM	4–8 PM	8 PM–12	12–4 AM	4–8 AM
Product	423 J				403 L								423 L	
TAKT (Rolls)	13	13	13	13	11	13	13	13	13	13	13	13	11	13
Produced (Rolls)	14	15	12	11	10	13	12	14	14	13	11	15	11	13
Ahead of TAKT	×	×	×											
Even with TAKT				×					×	×		×	×	×
Behind TAKT					×	×	×	×			×			
Reasons	*Ran well*				*tough changeover*						*drive failed*			
		Ran well					*paper tear-out*					*catching up*		
			adjust to takt					*catching up*						
				adjust to takt					*catching up*					

Figure 15.1 Example of a visual display board—A takt board.

ACTION REGISTER ~ Tuesday June 2nd
wobbling on the #3 idler roll - Fred
replace idler roll bearings - Ralph - noon today
drive on web winder failed
Replace drive - Wendy - done
Scraping noise when windup rolls transfer - Fred
check windup mechanism - Ralph - next product change - wed
Grease spot on floor below unwind roll - Jack
find source of grease - Ralph - before shift end today

Figure 15.2 Example of an action register.

- Allows recording of reasons for deviations from the plan
- Provides guidance on when to speed up or slow down the process to level production to takt
- Provides a communication tool between operators and maintenance
- Provides a communication tool between operators and supervisors
- Is therefore a very effective tool for managing the operation

Takt boards or similar visual displays are an important component of visual management systems, and thus a necessary part of any Lean production system.

The takt board is sometimes accompanied by a board or an easel with a flip chart pad to list all problems discovered, what follow-up action is to be taken, who is to do it, and when it is to be completed. Figure 15.2 gives an example.

For takt boards to be useful and effective, the operators who are the primary users must be involved in their design. The best boards are ones that have been designed by a team including operators, supervisors, mechanics, lab technicians, production schedulers, and anyone else who touches that part of the process. And for complete acceptance, all shifts should be given a chance to provide input.

Here are some of the best practices for scheduling board design:

- All information displayed must be relevant to the operation.
- Information must be clear and understandable to all users.
- Maximum use of symbols and graphics should be made to minimize text.
- The information must be updated frequently (somewhere between every hour and every 4 hours).
- The board should form the basis for the managing process.

Summary

No matter how good the product wheel design is, it will never happen until the various processes used to schedule the equipment are updated with the wheel strategy and operating parameters. This includes processes executed within ERP systems, manual input systems based on Excel, or completely manual processes. This also includes the visual systems used on the plant floor to inform operators what the current schedule is and where they should be on it.

Chapter 16

Step 13: Develop an Implementation Plan

With the wheel design now completed and approved, and visual board design and all computer system modifications underway, the next step is to put a plan together to make sure that all of the required tasks get done and to set a date to start the wheel.

Normal project management best practices should be followed here:

- Develop a list of all tasks required for start-up.
- Assign resources to each task.
- Determine the time each task will take.
- Set a target date for each task.
- Set a target date for overall completion.
- Have regular audits to make sure tasks are proceeding on schedule.
- Respond quickly to delays or potential delays.

Figure 16.1 shows a sample implementation plan. This list is shown simply as an example. There may be other steps required in your particular situation.

Task No.	*Task*	*Target Date*	*Responsibility*	*Status (as of June 11)*
1	Draft letter to senior management on make-to-order product recommendations	May 22	PJ + PK + MW	Done
2	Management approval for make-to-order recommendations	June 4	PJ	Done
3	Orientation presentation for sales and marketing, customer service	June 18	PJ + PK	Scheduled
4	Operator training—visual management, product wheels, their role	June 15	DS + PK	Scheduled
5	Design, fabricate takt boards and begin to use	June 22	LS + DS	
6	Write procedure to cover operators' responsibilities on visual boards	June 29	DS	
7	Define communications process for raw material supply to manufacturing line	June 15	DS + RH	
8	Design and implement a visual management board for raw material area	June 15	DS + RH	
9	Finalize all operating rules and procedures	June 15	DS	
10	Get approval for procedures	June 29	DS	
11	Complete all computer system configurations and master data changes and test	July 9	DR	Underway
12	Build inventories to cycle stock + safety stock targets	July 9	JM	Underway
13	Formulate and approve contingency plans	July 9	PJ + MW + DS	
14	Publicize the PW start date	July 9	PJ	
15	Allocate the first cycle PIT time	July 13	MW	
16	Start to operate the PW	July 16	PJ + JM	

JM = Janice Moore = production scheduler	DR = Doug Rash = senior programmer
MW = Mike Warren = production planner	PJ = Pamela Johnson = plant manager
RH = Randy Heck = raw material coordinator	LS = Larry Smith = lead operator
DS = Don Sellers = manufacturing engineer	PK = Pete King = Lean coach

Figure 16.1 Example of an implementation plan.

Chapter 17

Step 14: Develop a Contingency Plan

Even with the best product wheel design, situations that make people want to break into the wheel will occur. By breaking into the wheel we mean deviating from the sequence of products or the quantity to be produced on the current cycle, and producing something not scheduled in order to alleviate some critical issue. "Life comes at you fast," as the saying goes. It is important to anticipate all of the reasonably likely or somewhat likely problems so that appropriate responses can be developed before they will actually be needed. If most of these events and failures can be foreseen and planned for, it will go a long way toward preventing unnecessary wheel breakage, and minimize the negative impact when the wheel must be broken. It will also eliminate most of the stress, indecision, and arguing that can accompany these events, as well as the second-guessing and finger pointing that often occur afterwards.

While breaking into the wheel is never a good thing, there are circumstances that make that the best course of action. Good contingency planning will minimize the number of those circumstances, and enable the most appropriate plan of action to be followed when they do occur. This contingency planning is similar to the Six Sigma process of failure mode and effect analysis (FMEA).

Possible Wheel Breakers

There can be a number of situations that can lead to deviation from the planned wheel cycle. Among them are:

- Lack of a required raw material (or material from the previous step)
- Unexpectedly high demand for a product
- A management order to supply a make-to-order product within the make-to-stock lead time

- Mechanical and electrical failures
- Contaminated input material
- Poor performance on a very difficult-to-run product

Steps in Contingency Planning

1. Try to anticipate all of the serious problems that can reasonably be expected to occur. Use items from the list above as thought provokers.
2. For each one, decide if it is a valid reason to break the wheel.
3. Specify how bad it has to get (for example, how much of the abnormal demand will not be able to be supplied, how long it is expected to take to fix an equipment failure, etc.) for it to be appropriate to break the wheel.
4. Specify the course of action to be taken and the steps to get back on the wheel schedule.
5. Get agreement from all who have a voice in how to deal with these exceptional circumstances.

Example of a Contingency Plan

This is taken from a process that manufactures industrial lubricants and packages them in different container sizes. The product wheel sequence is optimized for the fewest fluid type changes, and secondarily for the fewest bottle size changes.

In this case, the leadership felt that the best way to handle emergencies was to sacrifice as much process improvement time (PIT time) as necessary to resolve the problem, and take further actions described below if that is not enough time. That may or may not be the best strategy in any specific situation, so this example is being provided as a sample of a contingency plan, but not to suggest that the specific actions should be used as a model.

POSSIBLE EVENTS AND REACTION

1. An unexpectedly high quantity order for a product (well beyond the normal cycle stock level):
   a. See if the customer is willing to take partial shipments. If so, phase the increased demand into the next few cycles. If not, go to 1.b. Also find out the reason for the large order—will this be a repeating situation? If so, adjust the cycle stock target for this product accordingly.
   b. If the wheel is currently on the spoke for that product (fluid and bottle size), enlarge the current spoke to make the normal amount plus the large order, if the time required can be regained at the expense of PIT time. If not, go to 1.f.

c. If the wheel is currently on the spoke for that fluid, but not that specific package size, finish the current product, then go to the added production when it makes sense within bottle sizes, etc. If this requires more time than available from PIT time, go to 1.f.
d. If the wheel is currently on a different fluid spoke, wait until the needed spoke comes around again—incur the stock-out (requires management approval). If management won't accept the stock-out, and if the product is very similar to a fluid being run, fit it in at a scheduled fluid change. If not similar to the current fluid, fit it in to the current wheel where it will incur the easiest fluid change. Absorb the extra time during the next PIT time period.
e. If it is a trigger point product (remember from Chapter 2 that a trigger point product is one that is made to stock, but not on any predetermined wheel frequency; it is scheduled to be produced on the wheel only when its inventory drops below a threshold, called the trigger point), the order will cause the trigger point to be hit. Make that product on the wheel the next time that fluid comes up. Reset to normal wheel schedule by shortening next PIT time period.
f. If the additional production requires more time than is available by sacrificing PIT time, incur the stock-out.

2. Shortage of materials from previous process step (in this case, this was not the first step in the process; it was fed semifinished material from the prior steps):
   a. Put in an urgent request for the semifinished material to be made. Skip that spoke. Insert the skipped spoke into the wheel at the next fluid changeover after the semifinished material becomes available.
3. Shortage of raw materials:
   a. Skip the affected spokes on the wheel. Put in an urgent request for the semifinished fluid to be made as soon as the raw material arrives. Once the fluid is available:
      1. If stocked out on any affected products, schedule them to be made at the next fluid change.
      2. If not stocked out, wait until that spoke comes up again.
      3. If not stocked out, but will become stocked out before the spoke comes up again, revert to 3.a.1.
4. Major equipment failure:
   a. Diagnose the failure. Order parts if not on hand. Repair failures as soon as parts are available. Resume wheel from point of failure. Skip the PIT time on this cycle, and on the next cycle if necessary. Reset to normal wheel schedule as soon as possible, which should be within two cycles. (The belief was that any part could be obtained and installed within 2 days. There was a day of PIT time on each cycle, so it was likely that the wheel could resume within two cycles.)

5. Fall behind takt for any minor upset or jam:
   a. Continue operation at standard rate. Complete packaging of the target quantity. Reset to normal wheel schedule by shortening next PIT time period.

Summary

Breaking the wheel is never a good thing, but under some circumstances it is the best course of action. Anticipating problems and developing appropriate responses at this level of detail will prevent wheel breaking when the preplanned alternatives are sufficient, and reduce the impact and severity in situations where breaking the wheel is the most appropriate response. It also minimizes the after-the-fact finger pointing and second-guessing that often follows unplanned reactions.

Chapter 18

Step 15: Get All Inventories in Balance

Chapter 12 described how to calculate the inventory needed for each product being made on the wheel. Now it's necessary to get all inventories close to those levels before we start the wheel. Typically, before any adjustments are made, some products will have much higher inventories than required, while others will be lower than needed.

We don't need the full cycle stock plus safety stock quantity of any product before starting the wheel, only the portion of the cycle stock expected to be consumed from wheel start until the spoke for that product is reached, plus the safety stock. So if a product is made on day 5 of a 7-day wheel, we need 5 days of the 7-day cycle stock, plus the safety stock, at wheel start.

And if the wheel is not scheduled to start for some period, the expected consumption during that period must be taken into account. In this case, the required inventory for any product equals the stock expected to be consumed for the period between now and the start of the wheel, plus the cycle stock expected to be consumed from the wheel start until the spoke for that product is reached, plus the safety stock, minus any planned production for that product between now and wheel start-up. So, if we plan to start the wheel 10 days from now, and the spoke for this product comes up on day 5 of the wheel cycle, we need 15 days of inventory plus the required safety stock, minus any scheduled production. If that amount is currently in inventory, we don't need to make any more until the wheel starts. If we have less than that, we should plan to make up the difference, so that after the normal consumption during the 10 days before wheel start, we will have 5 days of cycle stock plus safety stock at the start of the first wheel cycle. It is very important that at least the required inventory be present at the start of wheel operation. Any less can result in stock-outs, confusion, and deterioration of confidence in the wheel concept. Therefore any shortfall must be made up before the wheel can start.

If, on the other hand, any product has more than required, no action is needed. The result will be that very little (or none) of that product will be made on the first cycle, and maybe the second and third cycles if inventory was far above the target. The spoke for this product will be skipped until the inventory drops below the target. When this happens, it is very important to avoid the temptation to run through the cycle more quickly and start the next cycle early; any excess time due to empty spokes should be treated as additional process improvement time (PIT time), so that we stay on the designed cycle.

Summary

In order for the wheel to have a smooth start-up, the inventory for each product must be at or above the required level. If the inventory for any product is higher than needed, natural wheel operation will bring it within desired limits within a few cycles. If the inventory is lower than needed, it must be brought up to the required level before starting the wheel.

Chapter 19

Step 16: Confirm Wheel Performance—Put an Auditing Process in Place

After wheels have been started and are in routine operation, it is very important to make sure that the wheel discipline is being followed, and that performance is within expectations. With regular audits, any deviations from expected behavior will be recognized and can be diagnosed and resolved.

Things that should be monitored include:

- Customer service: Is on-time delivery performance as expected? Are your actual stock-outs very infrequent, in line with the service factors used to calculate safety stock?
- Inventory levels: Are they following expected patterns? Are you dipping into safety stock frequently? Are you diving deep into the safety stock on occasion? Remember that you should be consuming some safety stock on about half the cycles for each product.
- How many times did you have to break into the wheel? Was the appropriate contingency plan followed? Did you get back on the wheel as soon as practical?
- Are the takt boards being used appropriately? Do they show the current wheel schedule? Is performance against the takt goal being recorded? Are failures and problems being recorded on the board?
- Do the takt boards show that you are consistently keeping up with the scheduled production? Are you very often ahead of schedule? Or do you chronically fall behind and have to eat into the process improvement time (PIT time)? Performance to plan is a very useful metric to help understand this. The specific calculation used can vary with the application; a common one is the percentage of time blocks (say 4 hours) where the takt quantity was produced. Another typical definition is the percentage of days where takt was produced.

- Is the PIT time available approximately equal to the design? Is it being used to good advantage?

The results of this ongoing auditing process should be used to fine-tune the design to improve performance, or to find root causes and make corrections if performance is well below expectations.

- If customer service levels are below goal, you may need to increase safety stock. An inventory profile showing how often and how deeply you fall into the safety stock zone will help clarify the severity of any problems.
- If you are consuming safety stock too frequently, the average demand values used to set cycle stock should be reevaluated. If you too often go deeply into safety stock, the premises on demand variability, forecast accuracy, and process stability used to calculate safety stock should be reexamined.
- Similarly, if you rarely dip into safety stock, your cycle stock is probably too high and should be reevaluated. If, when you do use safety stock, you don't use much of it, and never get close to a stock-out, the premises on demand variability, forecast accuracy, and process stability are probably overly pessimistic and should be reexamined.
- If you find that the wheel gets broken into frequently, this is potentially the most serious of all these issues, and should be carefully analyzed. Is the constant turmoil because those managing the process are too "trigger happy"? Do they lack faith in the wheel concept, and begin to panic at the first indication of a potential problem? Or is it that the manufacturing process itself is not stable or capable?
- If takt boards aren't being used as intended, perhaps better training on their purpose and use should be conducted. Or it may be that the basic board design is confusing or difficult to use. If that is found to be the reason, getting all of the users together for a redesign would be the appropriate action. Takt boards almost always get upgraded and tweaked as people get experience with them, but sometimes a major redesign is in order.
- If performance to plan is consistently low, equipment problems are the most likely cause. This could include major failures, but the more common issue is a high number of short-duration line stoppages due to imperfect setups, with guides and fixtures not perfectly aligned, and other seemingly minor issues. Or in film, sheet, and fiber processes, constant tears or breaks when running difficult products may be the cause.
- If performance to plan is consistently at 100% or higher, overall equipment effectiveness (OEE) and effective capacity are probably understated; the equipment capability is higher than our current premises. In that case, the OEE metrics and the effective capacity calculations should be revisited.

Figure 19.1 summarizes the factors to be monitored, problems they may uncover, and areas to analyze to find the root cause of the difficulty. (The

Parameter to Audit	*Evidence of a Problem*	*Causes for Poor Performance*	*Action*
Customer service	Poor on-time delivery	Understated demand variability	Update demand variability calculations
		Not replenishing safety stock	Better replenishment management
		Not enough cycle stock	Reexamine demand data, recalculate cycle stock
Inventory performance	Dipping into safety stock on far less than half the cycles	Too much cycle stock	Reexamine demand data—look for downward trends
	Dipping into safety stock on far more than half the cycles	Not enough cycle stock	Reexamine demand data—look for upward trends, seasonality
	Running out of inventory on far less than 5% of the cycles	Too much safety stock	Update demand variability calculations, recalculate safety stock
	Running out of inventory on far more than 5% of the cycles	Not enough safety stock	Update demand variability calculations, recalculate safety stock
		Not replenishing safety stock	Better replenishment management
Wheel discipline	Frequent wheel breaking: not following sequence, not making correct quantity	Poor operating discipline	Emphasize importance of following the wheel to all involved
		Poor equipment performance	Improve maintenance; recalculate OEE and adjust capacity premises
		Inadequate contingency planning	Develop more robust contingency plans
Takt board discipline	The current wheel plan not shown	Lack of understanding of importance of boards and visual management concepts	Better training on visual management principles
	Performance to plan not being recorded		Reexamine takt board design with current users
	Problems not being recorded		
Performance to plan	Low percentage of performance to plan	Equipment failures	Improve maintenance
		Frequent jams and upsets	Autonomous maintenance
	Performance to plan regularly exceeds 100%	OEE and effective capacity are understated	Recalculate OEE and effective capacity
PIT time availability	Actual PIT time is consistently less than designed	Equipment problems	Improve maintenance
		Understated product demand	Reexamine demand data, recalculate cycle stock

Figure 19.1 Product wheel performance auditing.

descriptions in the table assume a 95% service level goal, and that demand history and demand variability, not forecasts and forecast accuracy, are used to set cycle stock and safety stock values.)

Chapter 20

Step 17: Put a Plan in Place to Rebalance the Wheel Periodically

After a wheel has been operated for some period of time, it is often the case that conditions have changed enough that the wheel design parameters should be recalculated. There should be an explicit plan to monitor the wheel design input values so that shifts requiring a rebalancing will be recognized and acted upon, rather than leaving it to chance that they will be recognized in a timely fashion.

Whenever any of the following factors change significantly, wheel design should be revisited.

- Changeover times or losses have been reduced through the continuing use of SMED and other changeover improvement practices.
- Throughput has been improved by opening a bottleneck.
- Overall equipment effectiveness (OEE) factors have changed: throughput has been improved by improving the reliability of a key piece of equipment or by reducing process yield losses.
- Demand for a mature product has fallen off.
- Demand for a newer product has grown.
- Demand for a product has become much more volatile.
- Demand for a product has become more stable.
- A new product is being added to the wheel.
- An obsolete product has been dropped from the wheel.
- Manufacturing costs go up or down significantly, due to raw material prices, energy prices, or labor costs.

Some of the factors discussed in the previous chapter, on auditing wheel performance, may provide signs that the wheel needs to be rebalanced. Of course, changes in any of the above parameters should be recorded in the data boxes on the value stream map (VSM).

Parameter That Changed	*Item to Recalculate*	*Likely Results*
Changeover time reduced	Shortest wheel	More avail time, could shorten wheel More PIT time
Changeover losses reduced	EOQ	EOQ will suggest shorter wheels Lower inventories
Manufacturing cost decreased	EOQ	EOQ will suggest longer wheels (inventory less expensive) Higher inventories
Demand for a product increased	Takt, utilization, EOQ, shortest wheel, cycle stock	Cycle stock will increase Utilization will increase EOQ may suggest shorter wheel or higher frequency
Demand for a product decreased	Takt, utilization, make-to-order, EOQ, shortest wheel, cycle stock	Cycle stock will decrease Utilization will decrease EOQ may suggest lower frequency
Product became more volatile	Make to order, safety stock	Safety stock must increase May be a candidate for make to order
Product became less volatile	Safety stock	Safety stock will decrease
New product added to lineup	Sequence, shortest wheel	Additional changeover; may lengthen wheel, increase inventory Less PIT time
Product taken off lineup	Shortest wheel	More avail time, could shorten wheel and reduce inventory More PIT time
Yield or equipment reliability increased	OEE, effective capacity, utilization, shortest wheel, EOQ	More avail time, could shorten wheel More PIT time

Figure 20.1 Factors to trigger rebalancing the wheel.

Wheels tend to be self-balancing to some degree, so minor changes will be accommodated within wheel operation relatively smoothly. However, because wheels tend to be self-leveling and self-balancing, performance will remain satisfactory in many situations where rebalancing would be beneficial. You need to monitor conditions so that you won't miss opportunities to significantly improve wheel performance based on improved process performance.

Figure 20.1 summarizes the factors that may change, the specific wheel design parameters to recalculate, and the likely effect on wheel design.

Chapter 21

Prerequisites for Product Wheels

With all the benefits that wheels can bring to an operation, there is often an eagerness to get started on product wheel design and implementation immediately. But it's better to resist that temptation until some foundational things are well in place. Wheels will have a much greater likelihood of success if the process is ready for them.

Product wheels require a stable process with predictable performance. Perfection isn't required, but stability and predictability are; we must know what throughput and uptime we can depend on. Product wheel design will be easier, start-up more successful, and performance better if built on a solid foundation.

Foundational Elements

The TPS (Toyota Production System) house (Figure 21.1) is often used to illustrate how the various components of a Lean operation are integrated. It shows the elements that form the foundation and support pillars, with the roof symbolizing the goal or purpose of the enterprise. Much of this model applies to product wheels as well. The entire system is built on a foundation of operational stability, standard work, well-maintained equipment, and visual tools used to manage the process. It is all formed around well-trained, highly engaged people.

The basic things that should be in place, or in active development, before beginning product wheel application include the following.

A Highly Motivated, Well-Trained Workforce

Operators must be well trained in their jobs and have value for standard work. There should be a healthy, trusting relationship between the workforce and

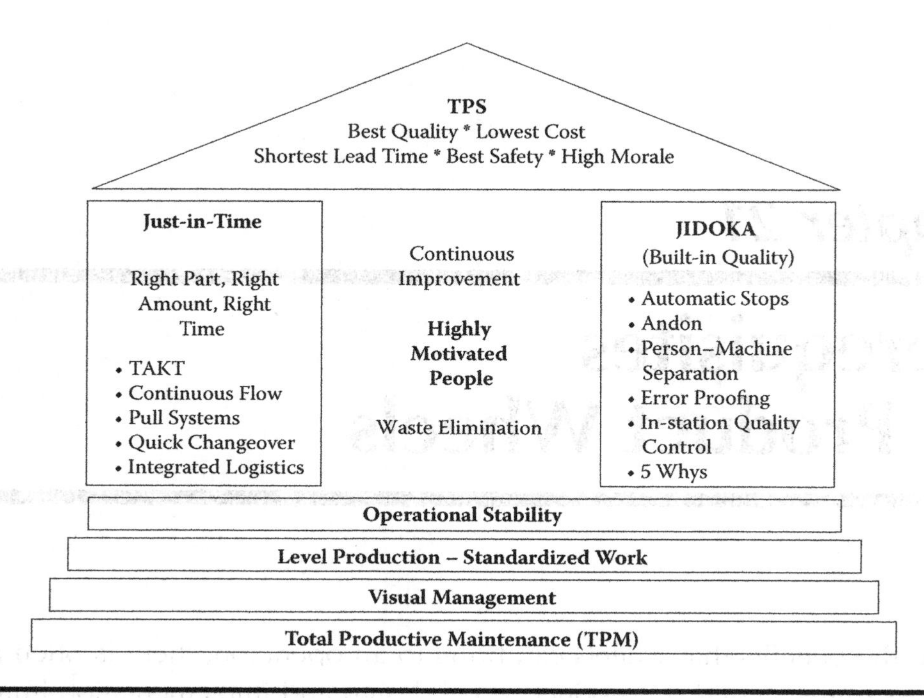

Figure 21.1 The TPS house.

management, and operators should feel engaged in the work and in improving it. Some understanding of Lean principles is also helpful, such as the nature of waste and the importance of finding and eliminating it.

Operators play a major role in product wheels, both in the design and in the ongoing operation. In many cases, they are responsible for taking the upcoming schedule and recording it on the takt board. They are the ones who must update the board with actual performance, and participate in decisions on how to proceed if performance falls short of takt. For these tasks to be done well, they need to know that they are a valued part of the operation, and feel a personal stake in its success. If the level of engagement is not all that it should be, getting operators, mechanics, and lab technicians involved in wheel design the way we've suggested will be a healthy step toward that.

Standard Work

All jobs must be standardized and documented. And then the standards must be routinely followed. Variability in the way the process is performed is as detrimental to smooth operation as variability in demand is. I've seen processes where operators were given a lot of latitude in how their job was done, and even though they were highly motivated, overall performance suffered when each shift did things very differently.

In one specific case, the crew working on a packaging line on the graveyard shift prided themselves in being able to run the line at higher rates than the other shifts. When I observed them one night, I found that they spent enough time getting the line set to the higher rate and tweaking all the adjustments, and then stopping to better align packaging film guides, that their overall production was less than that of the other shifts. If they had simply left all settings where they found them, the throughput would have been greater, and with fewer rejects.

If the line could indeed run stably at the higher rate, that should have been evaluated and verified, and then made the new standard. Adherence to standard work doesn't mean that improvements can't be made. It does mean that any potential improvements must be evaluated, demonstrated, proven, and then adopted as the new standard.

For product wheels to work effectively, everyone must follow the wheel rules and schedules, and not be allowed to deviate just because they think that there is a better way.

Visual Management

If visual management practices are not yet in place, product wheel implementation is a good place to start. The takt boards discussed in Chapter 15 allow operators and mechanics to become involved in design of visual management tools. The boards demonstrate the value of good, real-time, interactive visual displays, and if used properly, will become the basis for much of the day-to-day process management. Effective visual management will enable a much more timely response to any difficulties threatening wheel performance.

Total Productive Maintenance

In order for any product wheel scheduling to work effectively, the production equipment must run with a high degree of reliability and perform predictably. I have seen plants with excellent scheduling processes and very capable people executing them, yet be in total chaos. The culprit is almost always in equipment downtime, and not so much in hard failures, but in momentary upsets, like spills, overflows, and jams, because the equipment wasn't set up or adjusted properly.

I helped design a product wheel for a fluid blending and packaging line that had a stated reliability above 90%, but a detailed analysis revealed that only equipment failures were being recorded. When minor line interruptions were included, reliability dropped below 65%.

Total productive maintenance (TPM), and particularly the autonomous maintenance feature of TPM, can go a long way toward eliminating these short-duration but very frequent line stoppages. By giving operators the skills and tools to make adjustments properly and to fix minor problems, line uptime can increase dramatically. And their firsthand experience witnessing and fixing these problems makes them very well suited to help design the most effective longer-term corrective actions.

More reliable equipment will improve continuity of wheel operation, and also provide more freedom in product wheel design. Because reliability is a component of overall equipment effectiveness (OEE), increasing reliability increases OEE and therefore effective capacity. Thus less time is required to produce the required material, so more is available for changeovers, or for use as process improvement time (PIT time).

A more thorough description of TPM is given in Appendix C.

A Value Stream Map

The need for a good, up-to-date VSM was covered in Chapter 4, but it is important enough that a couple of key points will be emphasized here.

A well-constructed VSM will enable you to better understand process flow and scheduling issues, and therefore see where product wheels might be beneficial. Once the decision has been made to schedule a process step or an entire line by product wheels, the map will provide much of the data required in wheel design.

SMED

You should recognize by now that two of the most significant influences on wheel design are changeover times and changeover losses. Long changeover times reduce the time available for production, and can lead to longer wheels, particularly in equipment with high utilization. High changeover losses will drive the economic order quantity (EOQ) breakpoint toward longer wheels. Anytime you are faced with either long changeovers or high changeover costs, SMED (single-minute exchange of dies, a widely accepted method of analyzing and shortening changeovers) should be done, and repeated if necessary, to reduce time and losses. If SMED has been successfully done, resulting in shorter or less expensive changeovers, it will allow for a wider range of options in wheel design.

A more thorough description of SMED is given in Appendix D.

SKU Rationalization—Portfolio Management

Product wheel design is simpler, and has more latitude, if you only consider active, viable products. If there are a number of obsolete products, products with little or no sales, which remain in the lineup, they clutter up the wheel design spreadsheets and charts and mask opportunities for simplification. Extra products waste time in almost every step of the design process.

For that reason, all obsolete products should be weeded out of the lineup prior to wheel design. And for full effectiveness, there should be an ongoing process to analyze the portfolio and sunset products at some point in their decline, and not something that's just done once in preparation for wheel design.

One of the root causes of the problem is that many businesses have targets to gain a certain percentage of total revenue from new products, e.g., products introduced within the past year, or within the past 3 years. As these new products are introduced, there should be a process to reevaluate the existing products to weed out the ones that no longer make economic sense to carry in the portfolio. If SKU count is allowed to grow without limit, product wheels must handle an increasing number of product types, and manufacturing complexity, lead times, and inventories will grow while customer delivery performance may suffer. If a business has an active portfolio management process, production processes will be less complex and the performance targets easier to reach, and wheels will more nearly reach their full potential.

Bottleneck Identification and Management

Product wheels are much more effective if you have smooth continuous flow, with little interruption to flow. Therefore it is important to recognize if there are bottlenecks in the process flow, to open them up as much as possible, and then manage them so they won't inhibit flow any more than necessary. Opening any bottlenecks will increase capacity and provide more flexibility in making wheel design trade-offs, and can allow shorter wheels.

The VSM should clearly show any step that is a bottleneck or near bottleneck: its utilization number will be at or near 100%. Efforts to open the bottleneck should start with programs to improve OEE and its underlying factors, reliability and yield. Techniques to manage the bottleneck once it has been opened to the maximum practical amount are given in Appendix E.

Cellular Manufacturing and Group Technology

The application of cellular manufacturing was demonstrated in Chapters 1 and 2, where we took an array of four forming machines, four bonders, and three slitters and arranged them into four virtual work cells. This simplifies flow patterns, reduces the total number of flow path variations, improves quality, and makes the process easier to manage.

The greatest benefit comes when product families can be assigned to specific cells, so that each cell is making a subset of the full product lineup. That by itself will reduce the number of products to be scheduled on any product wheel, dramatically simplifying wheel design. But the real improvement comes from the fact that changeovers within a family are generally simpler and less costly than family-to-family changeovers, so a judicious allocation of products to cells usually results in the total cycle of changeovers on any piece of equipment being faster and less expensive.

The part of the cellular manufacturing design process that groups products into families and dedicates them to pieces of equipment or cells is sometimes called group technology. In cases where the process layout doesn't lend itself to virtual cells, group technology should still be considered where there are pieces of equipment in parallel. The reduction in the number of products run on any piece of equipment that results from group technology will simplify product wheel design, and much more so if the products can be grouped in a way that reduces changeover complexity.

Group technology and cellular manufacturing are discussed more thoroughly in Appendix F.

Summary

While product wheels can be designed and implemented without all of the above in place, design will be easier, implementation smoother, and performance much better if you have built a solid base on which to operate them. However, don't wait for perfection; if most of these prerequisites are in progress and moving toward success, the product wheel design/implementation process will reinforce their importance and accelerate their progress.

Chapter 22

Product Wheels and the Path to Pull

Product Wheels and Pull

As you develop product wheels in accordance with the methodology described in this book, you are actually implementing a pull replenishment system on the piece of equipment or line with the wheel.

Pull has two major components:

1. Replenishment of only what has been pulled from the downstream inventory (or what has been ordered, with a make-to-order or finish-to-order strategy)
2. Visual signals to convey to all involved what has been pulled from inventory (or ordered), to provide permission to produce new material to replace the pulled material

The methods we used to develop the wheel on forming machine 2 has satisfied both of these criteria:

1. By making the cycle stock plus safety stock minus the current inventory, rather than the designed cycle stock for a given spoke, we are in effect replacing only what has been pulled from the downstream inventory. Making the cycle stock amount without regard for the current inventory status is considered a push strategy, and can lead to overproduction and excess inventory. Push can also result in underproduction and stock-outs. Pull, on the other hand, restores the inventory to the designed level on each cycle.

2. The takt boards described in Chapter 15 are a very appropriate visual signaling mechanism. The computer system modifications described in that chapter provide an electronic signaling capability, and visual signals to those who have access to the appropriate forms on computer screens.

Figure 22.1 shows the cell 1 and 2 portion of the VSM, with forming 2 on a product wheel and the flow from forming 2 to bonder 2 on a pull replenishment strategy. The new icons on this map were previously defined in Figure 4.5. Inventory managed according to pull replenishment rules is often called a supermarket, because grocery supermarket shelves are generally restocked with what has been pulled from the shelf by customers. Thus the icon for inventory managed as pull is intended to look like a shelf. The oval-shaped arrow implies that bonding 2 pulls material from the supermarket as needed, and the thin arrow from forming 2 into the supermarket implies that material flows from forming 2 only in the quantity required to replace what bonder 2 has pulled. This is in contrast with the thicker cross-hatched arrows going from forming 1 to inventory, and to bonder 1, implying that this is scheduled by a forecast and is push replenishment.

To look more closely at how the pull system works, suppose that bonder 2 needs 12 rolls of product A to execute its schedule. Bonder 2 will pull 12 rolls from the supermarket, which will load a signal on the forming 2 wheel schedule to replace those 12 rolls. Suppose further that bonder 2 needs an additional 16 rolls of product A the next day to produce to its schedule. Again, it will pull the required number of rolls from the supermarket, which loads the requirement for 16 additional product A rolls on the forming wheel schedule. When the product A spoke comes around on forming 2, the number of rolls that have been pulled from the supermarket will be produced on the A spoke. Alternatively, forming 2 could simply wait until the spoke for product A arrives and then check the current inventory status to know what has been pulled and should be replenished. Each process gives the same result; the specific mechanism for accumulating the information is at the choice of the wheel design team.

Pull through the Entire Process

Once pull has been established and stabilized on one piece of equipment or one production line, it can be extended upstream and downstream to encompass the entire value stream. Figure 22.2 shows a portion of the sheet goods value stream map (VSM), at some time in the future when the entire process is on a pull strategy.

Raw material supply is now on pull. In the past, orders were placed with suppliers based on the monthly forecast. (We learned this from the information flow on the current state VSM shown in Figure 4.2) We now place raw material orders only when the inventory for any of the six materials drops below a trigger point. The trigger point is calculated so that we have enough inventory of that material left to continue to supply the process during the delivery lead time, plus some

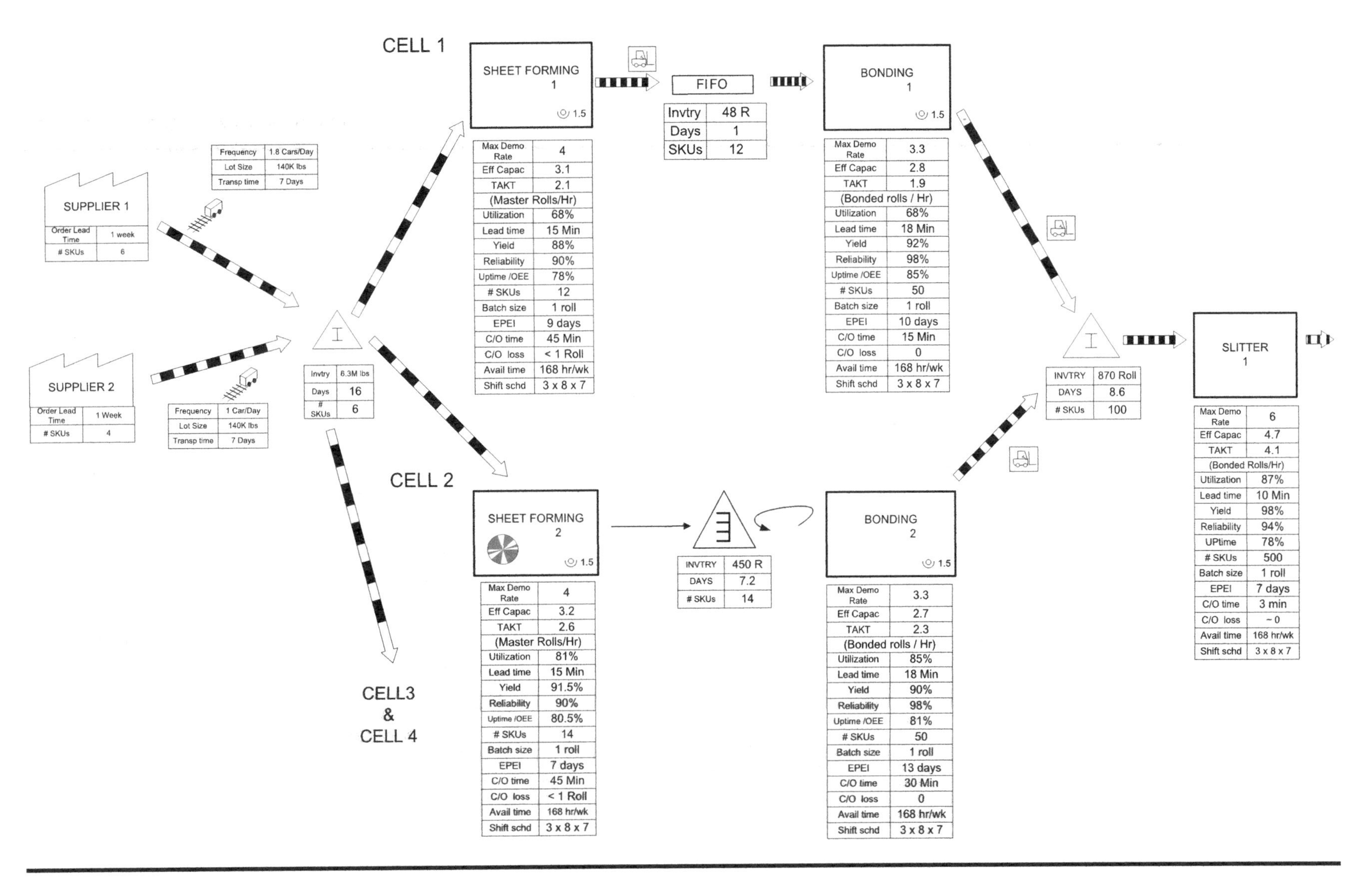

Figure 22.1 Flow from forming 2 to bonder 2 now on pull.

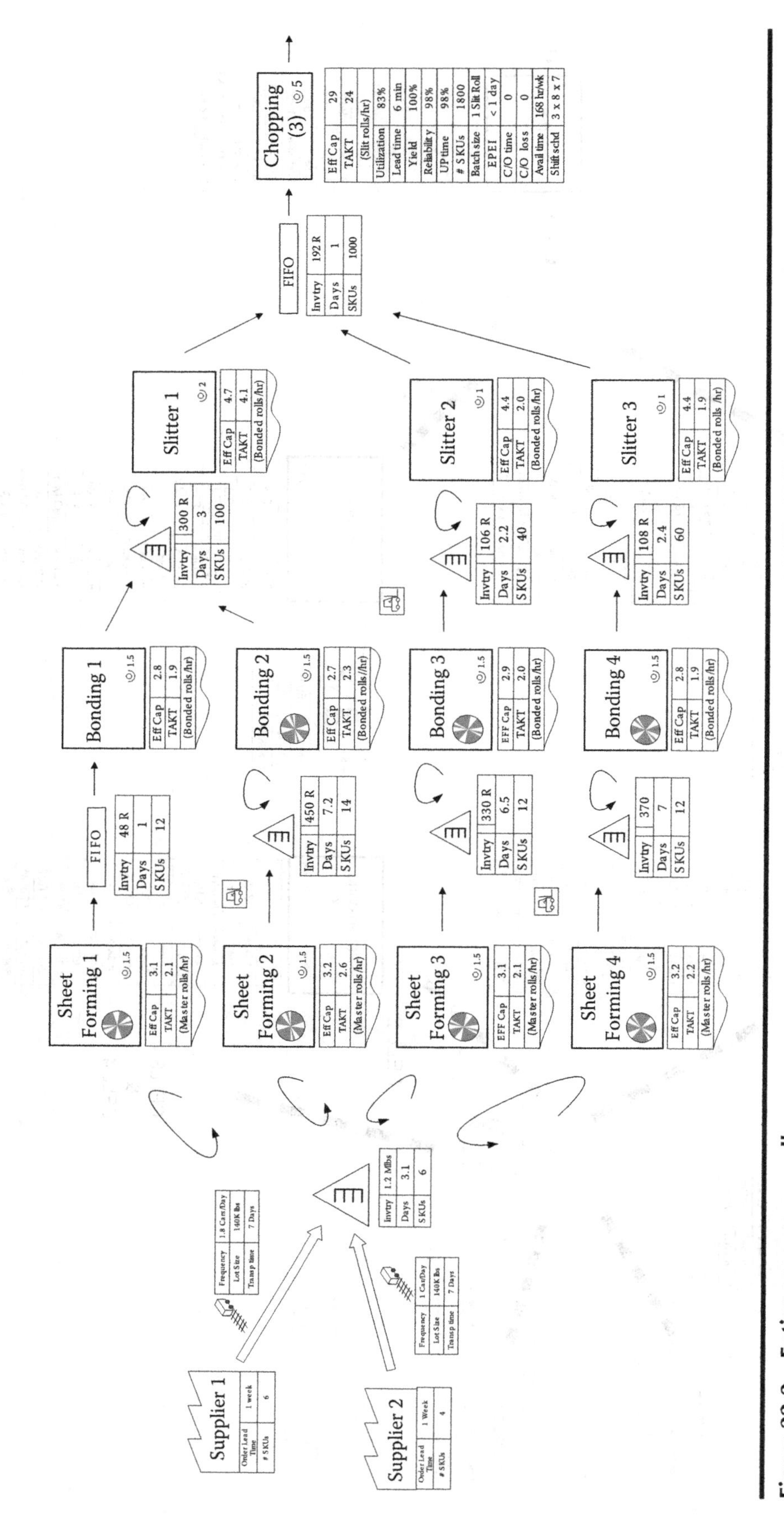

Figure 22.2 Entire process on pull.

safety stock in case the next truck arrives late. So this is a pull system in that we place raw material orders only when enough material has been pulled from the raw material supermarket to justify another order. This pull loop results in a very significant raw material inventory reduction, from 16 days to 3.1 days.

All four forming machines are now on product wheels, as are three of the bonders. You may remember from Chapter 5 that the product assignments done in the group technology step of the cellular design resulted in the group of products assigned to cell 1 having the same optimum sequence on forming and bonding, so we don't need individual wheels on both. (Group technology and cellular flow concepts are described in Appendix F.) When the spoke for product W is being run on forming 1, for example, that group of rolls is all bonded at the same temperature, so they can flow directly to bonding without requiring any inventory for regrouping. We do need some buffer between forming 1 and bonding 1 to manage the slight rate differences in the two machines, but that buffer can be much less than 1 day's inventory. It will be managed as a first in–first out (FIFO) inventory, and each roll will flow to bonder 1 as soon as the bonder has completed the previous roll.

Forming 2 and bonding 2 require separate wheels, because each product formed may have variants bonded at different temperatures. For example, product A from forming 2 must be bonded at four different temperatures to make bonded products A1, A2, A3, and A4. Similarly, product B from forming 2 must be bonded at three temperatures to make B1, B2, and B3. Therefore we need inventory between forming and bonding so that A1, B1, C1, and D1, which are all bonded at the same temperature, can be grouped together as one spoke on the bonder wheel. Products A2, B2, C2, D2, and E2 all bond at the same temperature, so they are grouped together as another spoke on the bonder wheel.

We had decided earlier that we didn't need wheels on the slitters, because the change in slitting pattern can be done very quickly, in 3 minutes, and because there is no preferred sequence. Although we can change slit pattern quickly, we don't want to do it on every roll processed, so it is desirable to have some inventory upstream of each slitter so that some degree of grouping of products with like slit patterns can be done. These inventories will be managed as supermarkets, in that bonding only replaces what slitting has pulled, but they can be small because each slitter runs relatively short campaigns.

So, with virtual cells in place, and with wheels implemented where appropriate, moving to a full pull replenishment strategy was straightforward. At that point, implementing pull was a matter of implementing the appropriate inventory managing practices. The inventories in a pull system are managed as supermarkets where there is a need to store material for some time to allow regrouping of product types to have similar processing requirements at the next step, or where replenishment is on a cyclic basis, and as FIFOs where neither of those is true.

Comparing Figure 22.2 to the original VSM, Figure 4.2, we see that the inventory between forming and bonding has been reduced by more than 50%, from 2500 rolls to 1198 rolls. Similarly, the preslitting inventory has been cut from 2400 rolls to 514 rolls, a 78% reduction! Total inventory on the plant, from raw material storage through finished products stored awaiting orders and shipping, has been reduced from 88 days to 26 days!

These reductions came about because of several factors:

- The application of rigorous mathematical tools to calculate true inventory requirements, instead of intuition, gut feel, or a need for the vague sense of security that extra inventory sometimes provides
- The flow simplifications and product assignments resulting from the virtual cellular manufacturing lineup
- The reduced campaign lengths developed in product wheel design
- Implementation of a pull inventory replenishment strategy
- Use of FIFO where there is no need to deliberately hold inventory

Summary

Pull replenishment is a key objective of any Lean initiative. A pull system schedules production based on current inventory status, and replaces only what has been consumed, rather than on a forecast that can become inaccurate very quickly. In that way pull avoids overproduction and leads to lower inventories. But even though inventories are generally lower with pull, the method increases the likelihood that the inventory is in those products most needed to satisfy customer orders, and so customer service usually increases.

While pull can be implemented at any time, it is easier to implement and more likely to give the desired results if group technology, virtual cells, and product wheels have been put in place wherever they are appropriate within the process flow. Virtual cells and group technology provide a sort of divide-and-conquer strategy, and product wheels optimize the product sequence and campaign lengths so that the inventories required by pull are at the lowest effective level.

By starting with the prerequisites described in Chapter 21, including virtual cells as appropriate, and then applying product wheels, you have built a very solid foundation on which to build a pull replenishment strategy.

Chapter 23

Unintended Consequences—Inappropriate Use of Metrics

Blind adherence to traditionally sacred manufacturing metrics can kill product wheel performance, so some of these metrics must be reevaluated in terms of what is important to the success of the wheels and therefore to the success of the operation. Measuring performance and comparing it to targets is generally a very good thing. It can highlight problems, give clues to root causes, and provide a basis to drive continuous improvement. However, unconditional allegiance to some normally good metrics can drive counterproductive behavior and quickly lead to wheel breakers.

Inappropriate Use of Metrics

Perhaps the most harmful practice involves a singular focus on productivity metrics, which measure production on a per-labor-hour or per-machine-hour basis or track costs on a per-unit-of-production basis. These drive manufacturing to run long campaigns to increase output per machine-hour or to lower the cost per unit. Product wheel performance requires that each spoke makes only the specific amount required, and not run longer just to enhance productivity numbers. Artificial productivity quotas generally drive overproduction, long campaigns, and excessive inventory. On a line with high utilization, productivity metrics often lead to wheel breaking.

Rather than measuring productivity, which rewards you for overproducing, it is much better to measure how well you're actually following the product wheel plan. Measuring how well you are meeting the schedule and delivering material to the downstream steps or to the final customer has much more value than measuring how much stuff you are making without regard to how much is actually needed. Performance to plan measures tend to drive production to the right levels, giving incentive to increase throughput when you're short, while not incentivizing overproduction when you're not. More on performance to plan (PTP) is given later in this chapter.

Overall equipment effectiveness (OEE; see Appendix C) is an example of a very good metric that can be misused to drive bad practices. If equipment is not fully utilized, if it is shut down for lack of demand for an appreciable percentage of available time, it may be appropriate to utilize that time to change products more frequently, to run smaller production campaigns and shorter wheels. This will cause more time to be spent in product changes, but that time is essentially free if the equipment is currently underutilized, and will result in reduced inventory, shorter lead times, and faster response to customer needs. However, because product changeover time counts against the OEE metric while idle time does not, and more of that idle time is now being spent in product changeovers, OEE will suffer even though a very beneficial change is being made. Those using OEE metrics as a gage of equipment performance must balance it against other operational metrics, but a conflict may arise when operational responsibility is stovepiped, with different individuals or groups accountable for OEE metrics and for inventory metrics. In many plants, the maintenance manager is responsible for OEE and will want to run long campaigns, while the operations manager is responsible for inventory and will want to run shorter campaigns, so there must be incentives for them to collaborate and do what is best for the operation's bottom line. Unilateral focus on OEE will always tend to drive fewer changeovers and longer wheels.

Another detrimental practice is focusing intently on cost reduction rather than waste reduction. Many businesses have aggressive cost reduction targets, and in pursuing them they remove necessary cost as well as unnecessary cost, reducing the operational capability of the process along with the waste. When this happens, the performance and indeed the health and stability of the operations deteriorate over time. This is like removing the muscle as well as the fat from the operation. If the focus could be redirected from cost reduction to waste elimination, as Lean teaches, the unnecessary cost would be removed but the necessary cost would remain. So the muscle, health, and stability of the operation would be intact.

Performance to Plan (PTP)

PTP should be focused on this question: "How much did I plan to make on this line (or specific process step) today, and how much did I actually make?"

Your product wheel design and current production plan are based on effective capacity, which takes OEE losses into account. Therefore you should expect to be able to stay on plan on average, to fall behind the plan when reliability, rate, and quality are worse than average, and to catch up when they are better than average.

The goal is to produce exactly to the production plan, and not overproduce; however, performance above plan is acceptable under two conditions:

1. You are behind the plan and are catching up.
2. You are on plan but overproducing to get ahead, knowing that OEE detractors, such as yield losses or equipment downtime, will hamper performance sometime in the near future. This is acceptable in the short term, but not over longer periods of time; you must always stop producing when the full takt for a spoke has been completed.

Therefore you shouldn't get penalized for producing above plan, but you shouldn't get credit either. Performance less than plan should be considered a detractor, and should be recorded as a negative. Performance greater than plan should be considered as being on plan if you are behind and catching up, or are getting ahead in anticipation of future problems, but only up to the total takt for any spoke.

There doesn't seem to be a standard definition for specific performance to plan metrics, so it's up to each operation to decide what makes most sense in terms of the parameters of the operation. One process for doing that comprises:

1. Decide what time periods make logical increments for collecting performance numbers. Each hour? Each 4-hour block? Each day? The answer will be different for different kinds of processes. For a fluid packaging line capable of 300 to 500 cases per hour, an hour or two may be a reasonable time increment for collecting performance data. For a sheet extruder capable of four rolls of film per hour, a 4-hour block may be best. For a chemical process with four reactors each running 3-hour batches, an 8-hour shift or a 24-hour day may be most appropriate.

   For a line scheduled by a product wheel, each spoke may be a good time increment, although for long spokes, a shorter time interval will generally give more useful insight.
2. Begin to collect data relating to each time block, and to calculate PTP on a regular basis.

$$\text{PTP} = \frac{\text{Actual production within a time block}}{\text{Planned production for that block}}$$

   (but not greater than 100%).
3. Plot the data, time block by time block, and use it to understand detractors, to drive improvement, and to gage whether you are improving.
4. It may be useful to also calculate performance on longer time blocks, say week by week, to understand performance over wider time spans. PTP should approach 100% as the measurement increment gets longer because overproduction to catch up will balance out periods of underperformance. Thus the primary benefit of calculating the value over large time blocks is to validate your OEE data and effective capacity numbers.

Summary

Metrics are almost always a good thing. It is important to have performance targets and to know where you are with respect to those targets. But the specific measures must be examined to make sure that they support the objectives and strategies of the business rather than unintentionally encouraging people to take counterproductive actions—actions that become wheel breakers.

Chapter 24

Cultural Transformation and Product Wheel Design—The Synergy

In the current business environment, almost everybody says that they want to change their culture, to be more inclusive, more open, and engage employees more fully in the objectives of the operation. But trying to effect a cultural transformation without some strong business case and specific improvement objectives to drive it is rarely effective. Culture change simply for the sake of culture change doesn't take hold very often. Similarly, trying to implement Lean processes, methods, and tools like product wheels rarely succeeds unless accompanied by changes to the culture, to the way that employees are engaged in improvement processes, and to the way that the people doing the value-adding work have a voice in determining how that work gets done.

Product wheels, if designed and implemented in the manner described in this handbook, bring the business benefit and the cultural benefit together in a very synergistic way.

- The primary purpose of product wheels is to improve operational performance in a way that improves business performance, through faster response to customer needs, improved customer service, lower inventories, and greater operational stability. The business case for product wheels is usually strong, and gives purpose and substance to the culture changes required.
- Product wheels require a high level of employee involvement in the design, initial implementation, fine-tuning, standardizing, and ongoing operation. For wheels to be effective, employees must feel that they are a valued part of the organization and have a stake in its success. As they move through all the steps we've described in this book, they must be confident that their voice will be heard, their knowledge and experience valued, their

ideas given serious consideration, and their efforts rewarded when success is achieved. Engaging line operators, mechanics, QC lab technicians, and others who are directly involved in the manufacturing process in the wheel design, and allowing their perspectives and ideas to have a strong influence on the choices to be made, will not only yield a better technical result, but also enhance their feeling of ownership of the results and their feeling of their value to the operation.

So product wheel implementation requires a high level of employee engagement, and to the extent that it doesn't already exist, wheel design provides a specific, substantial activity and a work process on which to build the engagement or to improve on what is there. The methods and practices recommended in this handbook provide a basis for getting operators, mechanics, technicians, and schedulers involved in a wide variety of activities: value stream map (VSM) creation and analysis, SMED changeover improvement events, implementation of virtual cells, product wheel design, and design and use of takt boards and other visual displays.

Another cultural benefit of the methods described in this handbook is that they will begin to move the plant away from a functionally oriented culture toward a flow-oriented culture. The paradigm in many process plants is to consider the operation in terms of functional areas: the polymer area, the extrusion area, the blending area, the packaging area, etc. Starting product wheel design with a good up-to-date VSM and using it as a design tool in the way described herein will shift the focus more along flow lines.

Summary

Unless a plant or an operating area has a very strong environment of employee engagement, trust, and open collaboration, it will have to be developed for product wheel design and operation to be successful. The good news is that the very methods used to design and operate wheels provide a platform on which to build that trust and collaboration. Wheel design provides a real and meaningful activity that provides opportunities to work on cultural improvement.

Almost any improvement to manufacturing work processes requires a high level of employee engagement to be sustained, and any improvement in employee trust and collaboration needs concrete, meaningful tasks to provide the basis for the increased involvement and responsibility. Thus there is an area of mutual need where culture improvement and wheel development can be mutually supportive and synergistic.

Chapter 25

Case Studies and Examples

Product wheels have been applied to a wide variety of processes, including automotive and house paints, extruded polymers, paper and plastic sheet goods, industrial chemicals, engine oil additives, waxes and pastes, and laminated circuit board materials. Here are a few specific examples to demonstrate the breadth of usage, and to illustrate some of the concepts covered throughout this book.

BG Products, Inc.—Automotive Fluids

BG Products is a major supplier of high-quality lubricants, brake fluids, and other engine performance-enhancing products to the automotive market. Located in Wichita, Kansas, BG Products has a successful performance record spanning more than 40 years. Because of the quality of BG products, the reputation of BG's brands, and a keen focus on customer service, BG Products has experienced strong growth, even in times of economic uncertainty.

Prior to the introduction of product wheels, BG had been able to satisfy customer needs at very high service levels, largely through the efforts of a number of highly dedicated, conscientious, experienced employees who took initiative to meet immediate needs. However, the lack of structured work processes and standard practices required constant monitoring of market and operational conditions to ensure continued high levels of customer performance. BG operations leadership recognized this and initiated programs to bring more structure and rigor to the operation, better definition of work practices, and higher adherence to work standards. Value stream mapping and product wheels were a key component of those programs.

The BG plant consists of a manufacturing area where the various automotive fluids are produced and approximately a dozen packaging lines each tailored to a specific range of product types and package configurations. One of the early improvement projects put product wheels on four of these lines. The first line addressed was the rotary filling line, where the fluid products were packaged into

6-, 12-, 32-, and 64-ounce bottles, then labeled, packed into cases, and palletized. Figure 25.1 shows a portion of the rotary filling line value stream map (VSM).

One of the early steps in wheel design was to reallocate products among the packaging lines to reduce the number of varieties packaged on any line. The 6- and 12-ounce bottles were moved from the rotary filling line to a line that generally packaged these smaller bottles, leaving only the 32- and 64-ounce bottles on the rotary filling line. There were 32- and 64-ounce products packaged on other lines, so they were moved to the rotary filling line. Certain fluid types were moved to a line where those fluids were more typically packaged. With fewer bottle sizes and fewer families of fluids, changeovers on the rotary filling line became faster and less expensive.

The next step was to examine order patterns and sales volumes for each product now to be packaged on rotary filling. We learned that there were quite a number of products that sold in very small volumes, small enough that it didn't make sense to carry inventory for long periods of time waiting for it to be sold. It was decided that any product with sales of eight pallets per year or less would no longer be made to stock, made only to fill specific orders, and that the minimum run for these products would have to be at least four pallets, the minimum practical campaign size. When the existing inventory of those products is worked down, it will significantly reduce total inventory. This make-to-order analysis was extended to all products made at the Wichita plant, resulting in about 130 products being identified, with a potential reduction in working capital of several hundred thousand dollars.

The product wheel sequence was designed to minimize fluid loss on changeover rather than time lost, because fluid loss represents a true out-of-pocket cost. Bottle size changes are time-consuming because they require changes to the filler valves, capper, conveyor guides, and case packer, but involve no additional costs. If the fluid type is changed, but packaged in the same bottle size, it can be done more quickly, but causes material to be lost in line flushing and filter cleaning, and so has a cost penalty.

The most extensive changeovers on the rotary filling line, with both fluid and bottle size changing, typically required 6 to 8 hours to complete, so SMED was applied. The initial SMED effort took the changeover down to less than 3 hours. To emphasize the importance of short changeovers and to drive further reduction, they continued to be tracked for some period, resulting in a changeover time below 2½ hours. Figure 25.2 is a sample of the chart used to track the more complex changeovers.

The optimum wheel length was determined to be 2 weeks, based on economic order quantity (EOQ) and minimum campaign size considerations. The variability in demand also had a strong impact on wheel length. Weekly demand for most products had very high variability; demand within 2-week intervals was much more uniform, with coefficients of variation (CVs) of much less than 0.5.

The current rotary filling line wheel includes 29 make-to-stock products: 10 of the higher-volume products are packaged on the 2-week cycle, while 8 others are

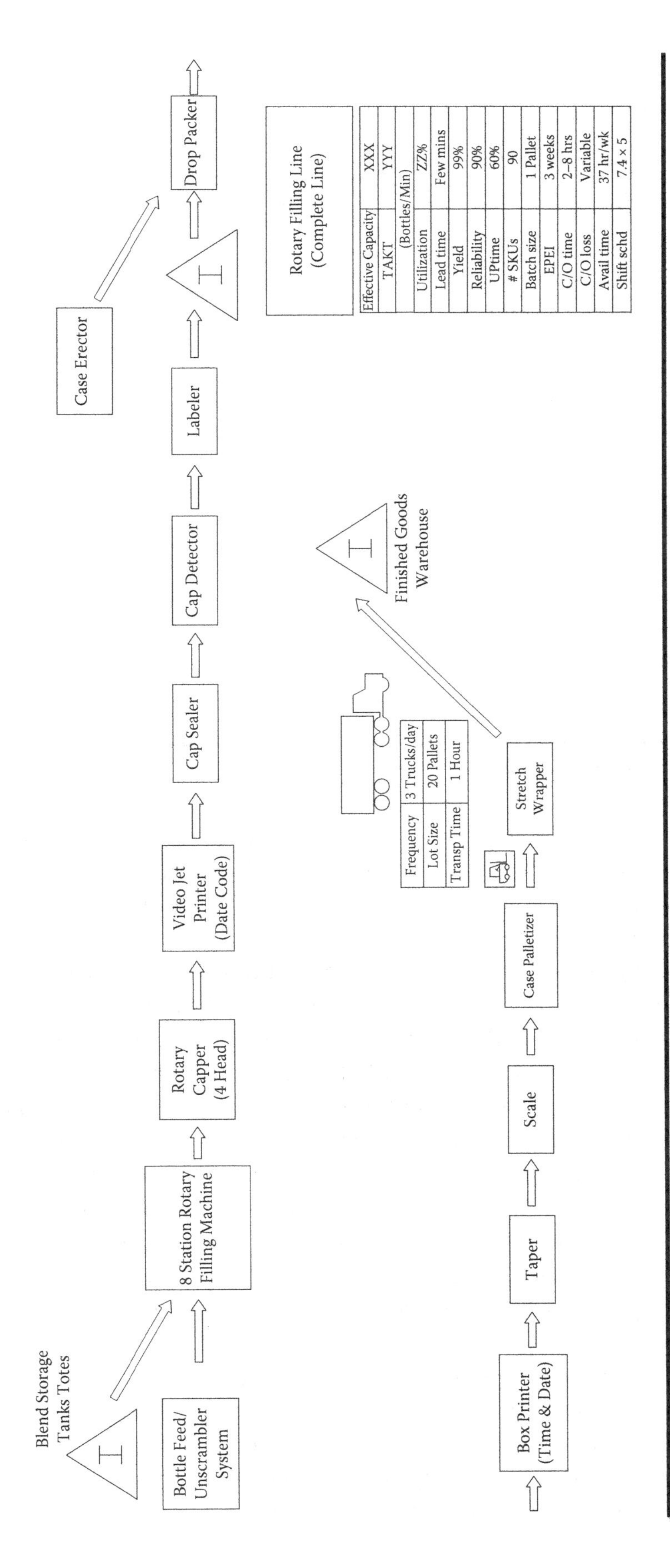

Figure 25.1 Rotary filling line current state VSM material flow.

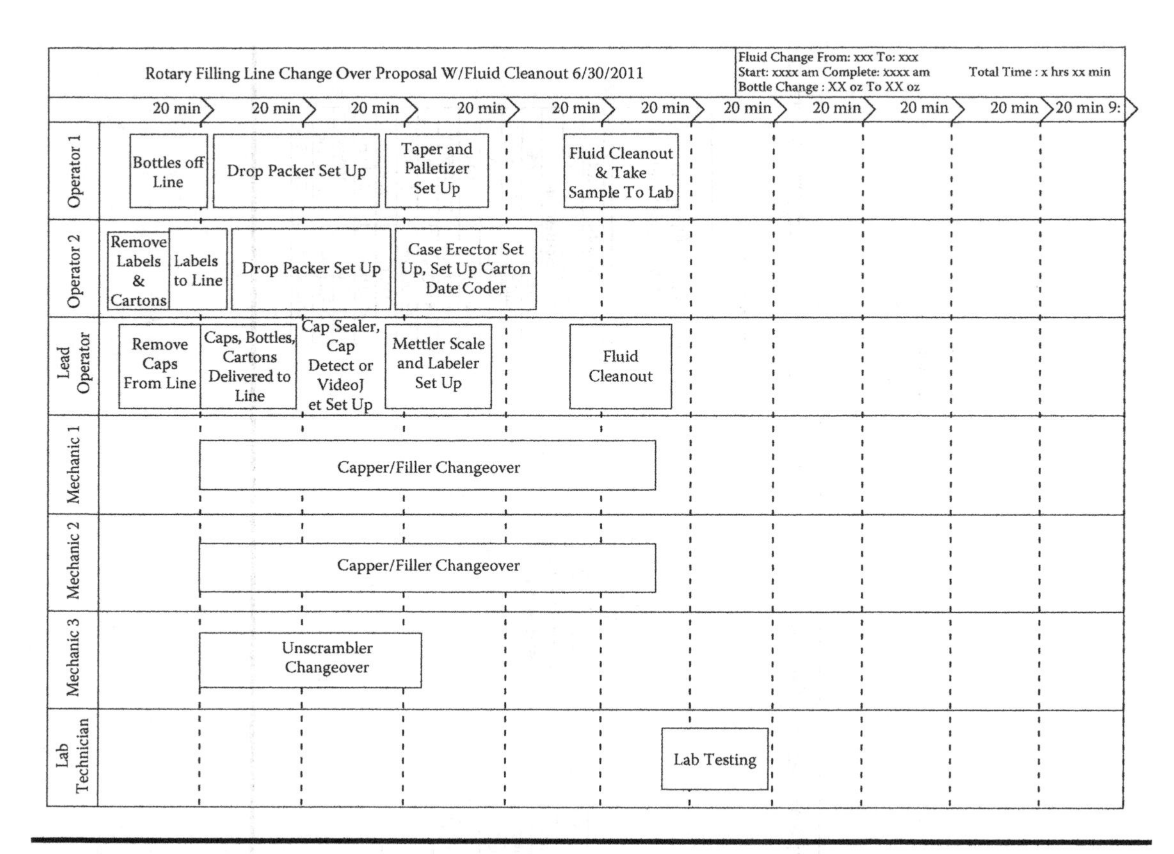

Figure 25.2 SMED tracking board.

made every second revolution. Eleven are packaged at a lower frequency, based on their forecast sales and the requirement for a minimum lot size of four skids per wheel spoke. There are about 25 very low-volume products, which are now made only to order.

Prior to beginning wheel operation, a team including lead operators designed takt boards for several of the packaging lines. The initial designs were similar to the board shown in Figure 15.1, and have been improved as operators and supervisors gained experience with the boards and a better understanding of how to use them.

The information being collected on the takt boards gave much greater insight into true line performance, and the reasons for some of the short but frequent line interruptions. In the words of the plant manager, "The takt boards are a key tool that indicates whether or not we are winning on that line on that day."

A key to the success of the wheel was the manufacturing management's decision to downplay the previously sacred productivity and line efficiency metrics, and to emphasize performance to plan measures, using the data being recorded on the takt boards.

Another key to the success was the highly interactive team effort, led and sponsored by the BG Operations Director, James Overheul. The cross-functional team included the production scheduler and planner, Lisette Walker and Matt Peterson, the plant manufacturing engineer, Gregg McCabe, with maintenance being represented by Dustin Mullen. Operator perspectives were provided by

Lorenzo Ruffin, Jonathon Blackwell, and Brent Scott. The plant management staff, including Jeremy Henry and Jeremy Lee, provided strong support and guidance throughout the implementation phase.

As a result of wheel implementation on the rotary filling line and three other packaging lines, BG has seen a significant reduction in inventories while continuing its very high levels of customer service. But perhaps the greater benefit has been the stability and predictability this has brought to the operation. Overheul summarizes his value for wheels as follows: "The process for creating the product wheels allowed us to see why we were having issues in our production environment. The product wheels gave us a process so that we can respond to the changing needs of our customers yet not lose our way on the routine items."

The Appleton Journey

Appleton is a specialty coating, formulation, and microencapsulation company headquartered in Appleton, Wisconsin. Serving global customers from four manufacturing sites, Appleton has become a market leader in thermal, carbonless, and security paper production, and microencapsulation technology. Appleton has a strong culture of continuous improvement, and has embraced Lean and product wheels as foundational elements of "The Appleton Way."

Appleton has been using wheel concepts very aggressively since 2008, and now has 11 wheels in successful operation. Ryan Scherer, Appleton's organizational excellence and capacity manager and the primary architect of its wheel design and implementation, describes Appleton's journey into wheel usage as follows:

> Appleton began working with the product wheel concept in the early 2000s, but it wasn't until 2008 that wheeling became a way of running the business. Appleton first implemented wheels on two of its carbonless coating machines to improve flow between these coating machines and the rewinders, and to reduce the volume of work in process (WIP). To deal with large production runs and the many inefficiencies and increased costs they created, wheels were employed as a structured way to reduce run length. Implementing wheels on the coaters allowed us to level load our large runs to eliminate the peaks and valleys in the schedule, and thus reduce overtime, create better flow, reduce WIP, and allow for a more predictable production schedule. And that was only the beginning.
>
> Observing the significant benefits achieved on the coaters, the carbonless value stream manager decided to apply the wheel concept to the sheeters. Many opportunities became apparent with this new way of thinking on how to schedule our equipment. Before the wheels were implemented, the same sheet parts would be running on two different sheeters at the same time, changeovers between cut sizes were

> very high, and finished goods inventory was out of control due to the unpredictable replenishment lead time. These difficulties were increased by the use of a forecast and "push scheduling." The sheeters were split into a "runner" and a "make-to-order/short run" sheeter, based on machine capabilities and design. Then using demand segmentation by SKU, wheels were built and sequenced appropriately to each dedicated sheeter. Again, this effort brought predictability to the schedule, significantly reduced cut size changeovers, and began to demonstrate a positive impact on finished goods inventory. At this point, momentum and interest in these improvements were growing across the business and other value streams.
>
> Because the results were so positive from the four wheels implemented on two coaters and two sheeters, our goal in 2009 was to get our three thermal coating machines on wheels. The thermal coaters were certainly more of a challenge than the carbonless coaters, but offered even greater opportunity due to high raw materials costs, and growing finished goods inventories due to the use of our ERP/MRP system and forecast mentality or "push" scheduling process. Again, we saw many of the results achieved earlier: improved schedule predictability, reduced and more efficient changeovers due to optimized sequencing, improved lead times, and fewer BSPs (broken service promises).

Scherer feels that it is the integration of three concepts (the "big three")—demand segmentation, product wheeling, and pull replenishment—that provides the key to dramatically improved performance.

Demand segmentation, analyzing each product by customer demand and by demand variability as measured by CV, described in Chapter 6, was the first opportunity to get the business team to stop thinking in terms of a forecast and start thinking about customers and their true demand.

The four segmentation quadrants (see Figure 6.4) were treated as follows:

Q1 = Runners. These "runner" parts made the biggest impact on finished goods inventory, as a result of level loading and reducing the run size. The runners are run every wheel cycle.
Q2 = Kanban. These are run every second or third cycle.
Q3 = Make to order. These are produced when orders are received, and are put into the schedule where similar runners and kanban parts are being run to reduce/eliminate schedule break-ins.
Q4 = Seasonal products.

Operator input was used to determine the optimum sequence of the various grades, being mindful of the eighth waste (lack of human potential, input, thinking). This resulted in the sheeter wheels being sequenced by roll width and then finish cut size, while the coaters are set up by roll width and then coating makeup and color.

All wheels were built and implemented using kaizen event methodology, to fully engage equipment operators and process engineers to include a wide range of knowledge and perspectives. Scherer feels strongly that data analysis is good to a point, but that shop floor knowledge is critical because the people actually doing the changeovers know changeover efficiencies the best. The kaizen team also included a production scheduler, inventory control analyst, and marketing specialist, with finance as an ad hoc member. This created a very strong cross-functional team with different perspectives on how to implement the wheels and how to get the full benefit from them, as well as buy-in from a change management standpoint.

Wheels are updated and reviewed at least once per year, but are dynamic and are adjusted as customer demand, production capability, and overall business strategy change.

There are now 11 machines on wheels: 2 sheeters (2008), 5 coaters (2008, 2009), and 4 paper machines (2010). All of the Appleton wheels, regardless of the type of equipment on which they were applied, have seen similar benefits: significantly lower inventories, shorter lead times, reduced changeover losses due to improved sequencing, and greater schedule stability and predictability, thus resulting in improved customer performance with fewer BSPs. The ongoing benefit of Appleton's Lean Six Sigma efforts, including product wheels and pull replenishment strategies, has been $20 million to $30 million annually, each year since 2008! Total inventory has been reduced by 21% and cash conversion days by 17%.

Being on product wheels has given Appleton the opportunity to work throughout its supply chain directly with customers, both domestically and internationally, by aligning wheels to customer demand so that customers see inventory reductions and increased service. When Appleton works with its customers on increasing supply chain efficiencies, it is always looking for win-win solutions.

Scherer sums up Appleton's value for wheels as follows: "I believe setting up and implementing production wheels is necessary, but is only the beginning. When implemented with a cross-functional team and used correctly, aligned with customer demand and integrated into inventory control (pull replenishment), wheeling can be very powerful."

DuPont™ Fluoropolymers

The DuPont™ Company has a successful global business in products based on fluorine chemistry, and several of the manufacturing sites use product wheels to schedule production.

One European site has five wheels in operation, with five to seven product types made on each line. Wheel design calculations led to a 10-day wheel time, versus the traditional 2- to 3-week campaign cycles. The calculations also validated the quarterly and semiannual production frequencies for particularly difficult products requiring extensive cleanouts.

Wheel lengths are reevaluated as business conditions change, in accordance with the principles discussed in Chapter 20. As expected, this resulted in shorter wheels during an economic downturn, and longer wheels as demand returned.

One of the domestic sites had a need to implement a wheel very quickly to resolve a serious customer service performance issue. The line being addressed produced 20 types, based on 5 reaction recipes. Forecasts had been used to plan production, resulting in high inventories of some products and stock-outs of others. This led to constant requests from the sales team for schedule changes to meet customers' short-term needs. Apparent sales-to-capacity ratios were near 100%, which gave little bandwidth to accommodate these frequent changes.

A plant team, guided by Lean Master Black Belt Rob Pinchot, adopted the principle of "don't let the perfect be the enemy of the good" to focus them on implementing an acceptable solution very quickly. Within a few days they had designed and implemented a wheel and pull system. After a transition period to reestablish inventories, stock-outs disappeared, the schedule became stable, and the only unplanned changes were in package type that had no impact on capacity. The elimination of the frequent illogical changeovers allowed available capacity to be utilized more effectively, with utilization dropping to a more practical level of about 85%.

As we've seen before, the structure, discipline, and predictability that wheels bring allow for both higher levels of customer service performance and lower inventories, in the absence of the churn and chaos that had typically been the norm.

Dow Chemical

The Dow Chemical Company is a $60 billion global enterprise serving specialty chemical, advanced material, agro science, plastics, and other industries. With a strong desire to become lean in order to improve flow and eliminate waste in its operations, Dow has made 5S, product wheels, and demand-driven pull systems a major focus of its continuous improvement initiatives. Dow applies wheels across all process areas, from the upstream continuous chemical reactions that may produce only a few products, to the downstream batch operations that convert those few materials into perhaps several dozen product variations. Today, Dow has 15 to 20% of its production facilities on wheels and pull.

Martin Fernandes, Dow's Director of Supply Chain Innovation, explains Dow's affinity for wheels thusly: "Wheels allow us to balance the tension between the desire to run long campaigns to minimize changeover difficulty and the desire for short runs to reduce inventories and to enhance flexibility to demand variation." Some of Dow's continuous chemical reactions can take more than a day to reach specified properties after a product change, which makes optimizing the sequence critical and campaign sizing a challenge. Fernandes uses computer modeling tools to assist Dow plants in balancing these variables while optimizing inventories to support pull replenishment processes. This application of wheels and demand-driven pull has resulted in 10 to 20% lower inventories, 30 to 40% shorter manufacturing lead times, greater operational stability and predictability, and more consistent lead times.

Fernandes concludes: "The greatest benefit is in delivery performance as seen by our customers, with fill rates 10 to 25% higher than with the old forecast-driven processes."

Extruded Polymers

A plant in Texas makes extruded polymer pellets sold to customers who mold them into various plastic parts. The plant has four extruders, and a lineup of more than 60 products that can be made on any extruder.

We began with extruder 1, and designed the wheel and the visual display boards during a 1-week kaizen event. Prior to the event, the plant process engineer applied group technology to the product lineup for extruders 1 and 2, which had shared 31 products, and was able to assign 15 and 16 products to extruders 1 and 2, respectively. The number of product families run on extruder 1 dropped from 8 to 2, dramatically simplifying changeovers.

The EOQ analysis and the other factors analyzed suggested that a 7-day wheel would be appropriate. Later kaizen events determined that 7 days was also optimum for the other three extruders, which was not surprising since the EOQ input parameters and other factors were very similar in each case.

Each wheel included a process improvement time (PIT time) of 24 hours or greater, and by staggering the start of the wheels, we were able to spread the PIT times over the week to smooth out requirements for maintenance and test laboratory resources. For example, extruder 1's wheel started on Tuesday morning, so Monday was available for maintenance, product qualification runs, and continuous improvement activities. Extruder 2's wheel started on Wednesday morning, so its PIT time came around on Tuesday.

The results were impressive. Extruder 1, for example, had been on a 21-day cycle, so the 7-day wheel allowed a significant inventory reduction, saving almost $400,000 in working capital. Even with the high inventory that had been available, the 21-day wheel had to be broken frequently to prevent stock-outs. Because the methodology described in previous chapters was followed to design the 7-day wheel, it was more robust and more stable, resulting in more predictability, far fewer wheel breakages, and far less frustration on the plant floor. And the easier changeovers from the group technology product lineup reduced yield losses by more than $2 million per year.

Waxes to Coat Cardboard

A global chemical company has a plant in Kentucky that produces waxes and oils used to coat corrugated boxes and cartons. The plant has two independent lines that make different grades of wax. Our focus was line 2, with a goal to find the optimum balance between inventory cost and changeover cost in hopes of reducing the current 28-day production cycle. Line 1 was considered to be completely independent of line 2, with its own set of dedicated products, so line 1 was out of the scope of the project.

However, when we constructed the VSM of the entire plant, we discovered that there was an earlier process in the plant that produced input materials for both lines. The map further clarified that while the input process generally had enough capacity to feed both lines simultaneously, there were a few products run on line 1 and a few on line 2 that consumed a large quantity of the input material. If both line 1 and line 2 were running products requiring a high flow from the input process, there was not enough capacity to produce that flow, and so one of the wax lines would be starved.

To remedy this possible constraint, we expanded the scope to include product wheels for both lines. We were able to coordinate the two wheels so that there would never be a time when both lines were on spokes requiring high-input flow.

This example reinforces the need to begin wheel design with a complete, accurate, up-to-date VSM, so that unusual situations like this will be recognized and accommodated.

Another interesting aspect of this application was that we were able to go to make to order on each line. On each line, there had been a 28-day cycle, and we determined that a 10-day wheel would be appropriate for each. While the 28 days was well beyond the 14-day customer lead time commitment, the reduction to 10 days put the manufacturing lead time within the customer lead time commitment, so make to order became feasible.

In some cases, a make-to-order strategy requires little or no inventory; you make and you ship. However, that was not the case here. While customers were satisfied with a 14-day lead time, they wouldn't accept early deliveries. They insisted on shipment on the requested dates. That meant that if the spoke for a particular product came around 6 days before the requested ship date, the product would have to be held for several days (6 days minus the lead time through the remainder of the process) before shipment. Thus the inventory required was approximately equal to the normal cycle stock, but the benefit was that no safety stock was required. On any spoke on the wheel, we knew exactly how much must be shipped on the next 10-day cycle, so there was no uncertainty of demand to be covered.

The design of both wheels, including the discovery of the limited capacity of the input material, occurred during a 1-week kaizen event. Preparation for the event took place over a 3-week period, and included the development of the VSM.

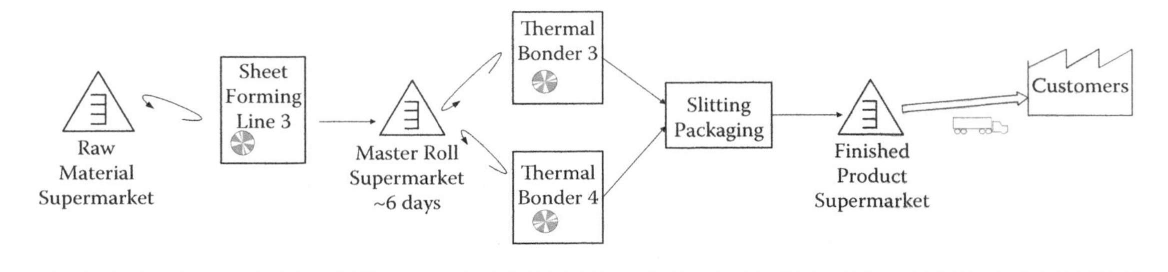

Figure 25.3 Sheet goods future state VSM 1.

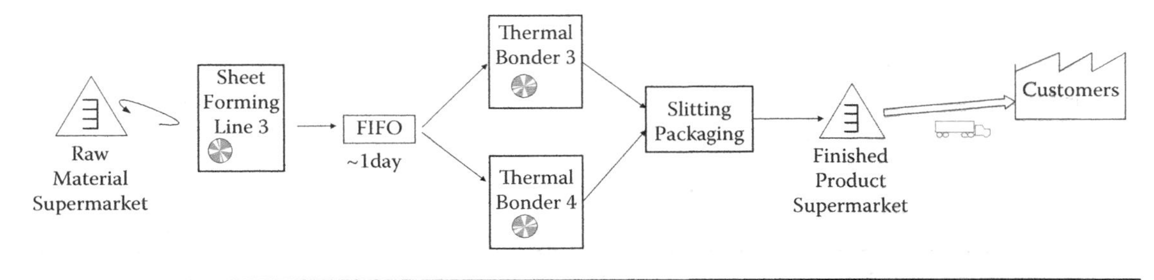

Figure 25.4 Sheet goods future state VSM 2.

Sheet Goods for Hospital Gowns

A plant on the East Coast of the United States makes a paper-like fabric that has several end uses. One of the higher-volume product families is used to make hospital gowns, curtains, and covers for operating room tables. The manufacturing process is very similar to the process used to illustrate the wheel design methodology throughout this book, with sheet forming, bonding, slitting, and packaging.

The initial wheel design resulted in the future state VSM shown in Figure 25.3. Wheels were designed for the forming line and the two bonders used to produce this product family, and pull replenishment was to be employed across the entire process.

Further analysis revealed that we didn't really need a supermarket between the forming machine and the bonders. The wheels on the two bonders could be synchronized so that the two were bonding a similar formed product at any point in time. And within this product family, forming changeovers are simple enough that the forming wheel could be synchronized with the bonder wheels, running approximately 24 hours ahead of the bonder wheels. So at any time, the exact amount needed for the next 24 to 48 hours of upcoming spokes on each bonder would be communicated to the forming scheduler so that the forming spoke amounts could be determined. The net result of this was that a supermarket of approximately 6 days of inventory was replaced with a FIFO (first in–first out) inventory of approximately 24 hours, as shown in Figure 25.4.

Circuit Board Substrates

A plant in western Pennsylvania produces laminated sheets of plastic film containing layers of conductive materials and insulating materials, from which

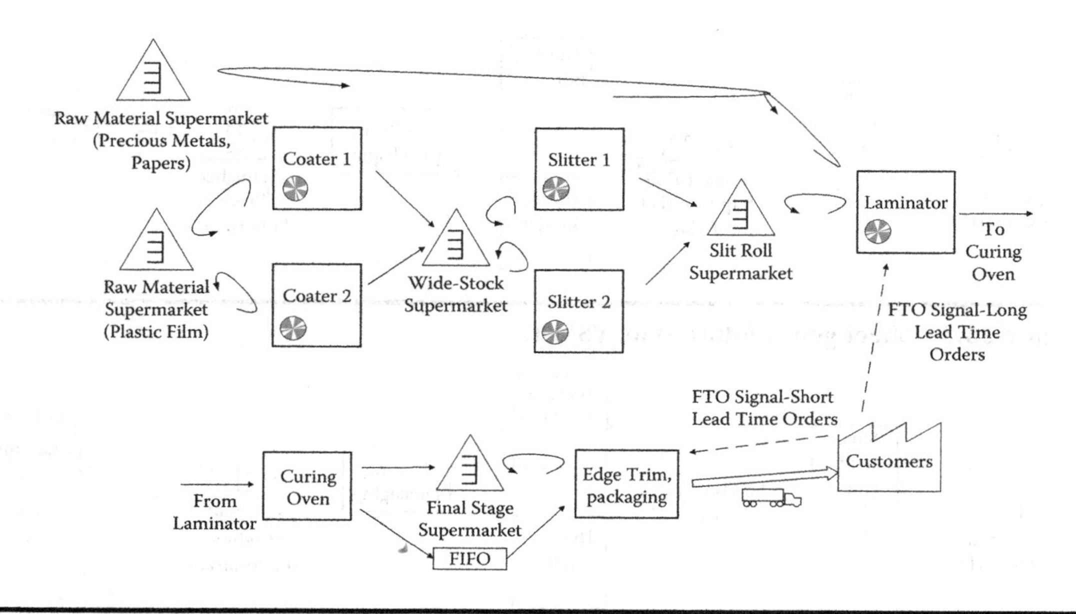

Figure 25.5 Circuit board substrate manufacturing process with product wheels and pull flow.

customers make flexible circuit boards. A very simplified version of the VSM reflecting the flow after application of product wheels and pull replenishment is shown in Figure 25.5.

Cellular flow couldn't be applied here. Coater 1 and coater 2 have somewhat different processing capabilities, as do slitter 1 and slitter 2. There are products that must be processed on coater 1 and on slitter 1, while others from coater 1 must go to slitter 2. Looking at the full product lineup, all four combinations of coater to slitter flow paths are needed.

You can see from the map that we have wheels in series and in parallel, which is not uncommon. What is unusual about this process is that we are finishing to order (FTO) from two points in the process flow. There are some products for which customers require shipment within 2 days of the order. The laminated and cured material for these products is stored in the final stage supermarket, so that when an order is received, the material can be trimmed, packaged, and shipped immediately. The material in the final stage supermarket is quite expensive, because it contains the precious metals, and so we must keep that inventory at a minimum.

Fortunately, most of the products have a much longer lead time commitment, so material to finish those products is stored in the slit roll supermarket. When an order for one of those products comes in, it gets loaded on the laminator wheel schedule. When that spoke comes around on the laminator wheel, the appropriate material is pulled from the slit roll supermarket, laminated, and cured. Rather than flowing into the final stage supermarket, it flows in FIFO fashion to edge trim and packaging.

Because this dual-FTO strategy is very unusual, a simulation model was developed to validate the practicality of the concept. The model was also helpful in fine-tuning the supermarket inventory levels, which was especially important for the expensive final stage inventory.

This design resulted in a reduction of in-process inventory of 40%. Customer service improved dramatically: where in the past, the average lead time for the final stage products was 7 days against the 2-day commitment, this production strategy cut it to 1 day. Lead time for the products finished from the slit roll supermarket was cut from 14 days to 10 days.

Fixed-Sequence Variable Volume

In *Liquid Lean*, Ray Floyd describes a production scheduling technique called fixed-sequence variable volume (FSVV), which is very similar to the product wheel methodology we've been describing. As its name implies, finding the optimum sequence to minimize changeover time and cost is of paramount importance, just as it is with wheels. Like wheels, FSVV follows pull replenishment principles, where what is produced during any campaign is based on actual consumption rather than on any predetermined amount. And like wheels, it produces the lower-volume products on a frequency less than every cycle.

The key difference is that rather than setting a fixed wheel time, FSVV allows total cycle time to float. The primary reason is that it has generally been used with complex chemical polymerization reactions that are extremely capacity limited. Therefore, with high demand and difficult changeovers, operations tend to run long cycles. The key to shortening the cycle time is reducing changeover time by finding a better sequence, but determining the optimum sequence is very difficult, and is therefore a focus of continuous improvement activities. Incremental sequence improvements are being made regularly, so effective capacity is almost continuously increasing while cycle time is almost continuously decreasing.

Although overall cycle time is allowed to float, it is consistent enough from cycle to cycle that cycle stock requirements can be determined reasonably well.

All in all, the two scheduling methodologies are far more alike than not, and offer similar benefits. Floyd reports that an Exxon polypropylene plant in Baytown, Texas, was able to reduce changeover losses by 90% and increase reactor effective capacity from 50% of nameplate to more than 85% using the FSVV methodology.

A Rose by Any Other Name ...

A major player in the pharmaceutical industry makes extensive use of product wheels, and refers to them as rhythm wheels.

Some companies in automotive paint and architectural paint production use product wheels, and refer to them as color wheels.

Summary

These examples should give you a feel for the wide range of processes that have benefitted from product wheel scheduling. And although the processes are quite different, the benefits have a lot in common: reduced inventories, increased usable capacity, shorter lead times, reduced changeover losses, higher customer service performance, and perhaps most importantly, greater stability, predictability, and reduced chaos. Stabilizing the schedule into a regular pattern allows the normal events to proceed with far less attention, so that more attention can be paid to the abnormal events.

Bibliography

Blackstone, John H. Jr. (Editor). *APICS Dictionary,* 12th edition. Chicago, IL: APICS, the Association for Operations Management, 2008.

Bowersox, Donald J. and David J. Closs. *Logistical Management: The Integrated Supply Chain Process.* New York: McGraw-Hill, 1996.

Chopra, Sunil and Peter Meindl. *Supply Chain Management: Strategy, Planning, and Operation.* Upper Saddle River, NJ: Pearson Prentice-Hall, 2007.

Floyd, Raymond C. *Liquid Lean.* New York: Productivity Press, 2010.

Goldratt, Eliyahu M. *Theory of Constraints.* Great Barrington, MA: North River Press, 1990.

Goldratt, Eliyahu M. and Jeff Cox. *The Goal.* Croton-on-Hudson, NY: North River Press, 1984.

King, Peter L. *Lean for the Process Industries: Dealing with Complexity.* New York: Productivity Press, 2009.

Nakajima, Seiichi. *Introduction to TPM.* Cambridge, MA; Productivity Press, 1988.

Ohno, Taiichi. *Toyota Production System: Beyond Large-Scale Production.* New York: Productivity Press. 1988.

Panchak, Patricia. Leveling and pull streamline production in process industries. *Target Magazine,* 25(1), 2009.

Rother, Mike and John Shook. *Learning to See: Value Stream Mapping to Create Value and Eliminate Muda.* Cambridge, MA: The Lean Enterprise Institute. 2003.

Shingo, Shigeo. *Quick Changeover for Operators: The SMED System.* New York: Productivity Press, 1996.

Shingo, Shigeo and Andrew P. Dillon. *A Revolution in Manufacturing: The SMED System.* Cambridge, MA: Productivity Press, 1985.

Silver, Edward A. and Rein Peterson. *Decision Systems for Inventory Management and Production Planning.* John Wiley & Sons, 1985.

Smith, Wayne K. *Time Out.* New York: John Wiley & Sons, 1998.

Umble, M. Michael and Mokshagundam L. Srikanth. *Synchronous Manufacturing.* North River Press, 1990.

Appendix A: Cycle Stock Concepts and Calculations

Inventory Components Defined—Cycle Stock and Safety Stock

Where inventory is being maintained to support a product wheel, it generally has two components, cycle stock and safety stock.

Cycle stock is the amount of a specific product to be made during the spoke for that product, to satisfy demand over the full product wheel cycle, until the spoke for that product comes around again. For example, if the wheel time is 7 days, the cycle stock for material X would be 7 days of average demand. If material X occupies 1 day on the wheel, at the end of its production day there must be 6 days of material in the finished goods warehouse, or in downstream process steps and headed for the warehouse. That material is needed to satisfy demand for product X in the 6-day interim until X will be made again. So the cycle stock for X includes the 1 day that was consumed while X was being produced and the 6 days to satisfy demand during the rest of the cycle.

The second component of inventory is safety stock, material held to satisfy demand in cases where actual demand is higher than expected, or where the next cycle was late in starting.

Figure A.1 shows a profile of inventory versus time for a single product, where cycle stock and safety stock are present. In production period P1, which is the spoke for that product, cycle stock is produced, to raise the level to A. Demand during the rest of that wheel cycle, D1, is equal to the average demand, so the cycle stock is consumed, but safety stock is not. Production P2 raises total inventory back to level A. Demand during the next cycle, D2, is higher than average, so that in addition to consuming all of the cycle stock, some of the safety stock is needed. This would also be the case if it took longer than average for the process to complete its cycle and return to making this material. Thus the safety stock will protect flow against either variation in demand or variation in supply lead time. With good product wheel management, variation in manufacturing lead time should be minimal, so variation in demand will be the primary factor requiring safety stock.

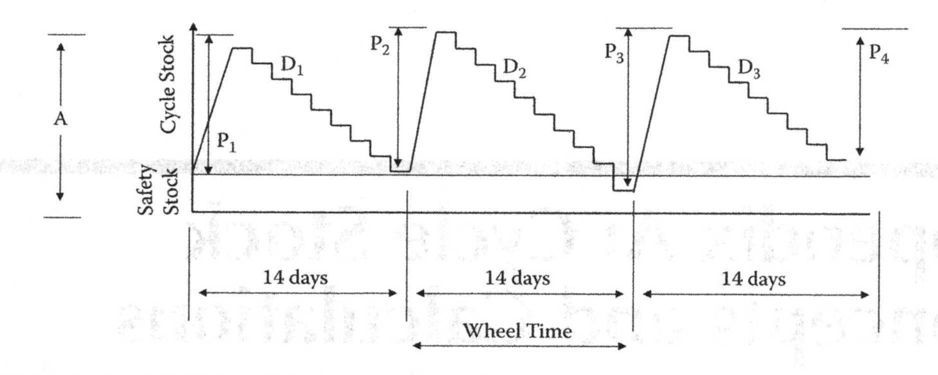

Figure A.1 Cycle stock and safety stock.

Production P3 must be greater than average, to replace cycle stock plus the amount of safety stock that was consumed. Production P4 is less because no safety stock, and not even all of the cycle stock, was consumed.

Cycle stock is based on the average demand expected. This can be based on either demand history or a forecast. If previous demand is considered to be the best predictor of future demand, then demand history should be used to set cycle stock. If there is a forecast that is believed to be a more accurate indication of future demand, then cycle stock should be based on the forecast. Since forecasts can vary period by period, the cycle stock may be adjusted upward or downward each period in accordance with the forecast.

Calculating Cycle Stock—Fixed-Interval Replenishment Model

Cycle stock can be replenished on a fixed-interval or fixed-quantity basis. Since product wheels are a fixed-interval strategy, that is the model we'll discuss here.

The inventory profile shown in Figure A.1 represents a fixed-interval strategy, with a 14-day wheel time. In this situation, we must make enough to last until the next production of this material, or 14 days' worth. So the cycle stock will be the average demand during a 14-day period, and the peak inventory will be cycle stock plus safety stock, or 14 days' worth plus safety stock, minus material consumed during that spoke. What actually gets produced when that spoke of the wheel comes around is not always the cycle stock, but depends on current inventory. We saw that demand D2 was slightly greater than average, so some of the safety stock was consumed, and the quantity P3 had to include the normal cycle stock plus the amount of safety stock that was consumed.

The equations that apply to those situations are:

$$\text{Peak Inventory} = \text{Cycle Stock}\left(1 - {}^{D}\!/_{PR}\right) + \text{Safety Stock}$$

$$\text{Average Inventory} = {}^{1}\!/_{2}\,(\text{Cycle Stock})\left(1 - {}^{D}\!/_{PR}\right) + \text{Safety Stock}$$

where D is the demand for that material per unit of time, and PR is the production rate, the total quantity produced over that same time.

The $(1 - D/R)$ term is there to reflect the fact that some of the cycle stock is being consumed by downstream steps during the production cycle. This has a very minor effect on products with a relatively small spoke on the wheel, but can be very significant if a product has a large spoke, i.e., occupies a large portion of the cycle.

The quantity to be produced on any cycle will be

$$\text{Quantity Produced} = \text{Cycle Stock} + \text{Safety Stock} - \text{Current Inventory}$$

Because the current inventory will on average be approximately equal to the safety stock, the quantity produced will generally be approximately the cycle stock.

The replenishment model used on product wheels is called fixed interval because the replenishments for any product occur at very regular intervals on a regularly repeating cycle, where the time between replenishments may change only slightly, but the quantity can change significantly, depending on how much material has been consumed during the most recent cycle. This is in contrast with the fixed-quantity model, which behaves exactly the opposite. In that model, the replenishment quantity is always fixed to a very specific amount, but the interval between replenishments can vary significantly.

This fixed-interval model is sometimes referred to as a fixed-order interval (FOI) model or a periodic review system. This is the same replenishment process as is used in a grocery supermarket where the shelves are restocked on some regular basis, say every Friday morning. The interval is fixed, Friday to Friday, but the quantity will vary based on the amount customers have pulled from the shelf since the previous Friday.

Summary

The inventory for any product being made on a wheel consists of cycle stock and safety stock. Cycle stock is determined by average demand, the length of the wheel, and the frequency for products not made on every cycle. Safety stock is there to protect against stock-outs in the face of variable supply, production, or demand. It is determined by the variables we want to protect against, which for product wheels is largely variability in customer demand.

Appendix B: Safety Stock Concepts and Calculations

About Safety Stock

Safety stock is inventory carried to prevent or reduce the frequency of stock-outs, and thus provide better service to customers. Safety stock can be used to accommodate:

- Variability in customer demand or in demand from downstream process steps (where demand history is used to set product wheel cycle stock)
- Forecast errors (where forecasts are used to set cycle stock targets)
- Variability in wheel time

If demand and wheel time vary randomly and are reasonably normally distributed, the following calculations will result in appropriate safety stock levels. If not, they may still give some guidance, and are generally preferable to the sometimes recommended rules of thumb, that safety stock be set at 10, 20, or 50% of cycle stock.

Variability in Demand

To understand how we can avoid stock-outs in the face of customer demand, which can vary up and down, a short lesson in statistics is in order. Figure B.1 is a histogram, a plot showing the number of cycles at which each demand range occurs. If we consider product A on forming machine 2, with an average demand of 130 rolls per weekly product wheel cycle, the histogram shows how many weeks the true demand was within each range. The histogram shows that, for the 52 wheel cycles within a year, the demand was very close to the average for 12 of those weeks. In this plot, the width of each bar represents 10 rolls; so on these 12 weeks, the demand was between 125 and 135 rolls. It was somewhat higher, 135 to 145 rolls, during 8 weeks, and between 145 and 155 rolls during

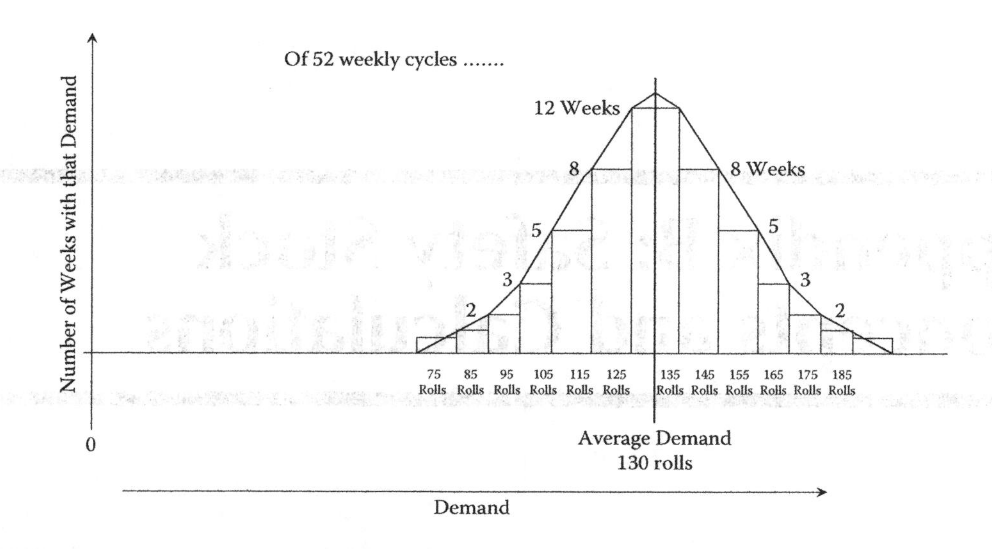

Figure B.1 A histogram of weekly demand.

5 weeks. As the range of demand values goes higher, the number of weeks within that range decreases. There is a similar pattern on the other side of the average; for 8 weeks, the demand was between 115 and 125 rolls, and between 105 and 115 for 5 weeks. This bell-shaped curve is typical of many demand patterns.

Some products will have little variability and thus a very narrow histogram, while others will have higher variability and a wider histogram. The width of the curve, and the underlying variability, can be characterized by a statistical property called standard deviation and symbolized by sigma, σ. While the calculation of standard deviation is beyond the scope of this discussion, understanding σ can help us calculate how much safety stock we need to give us various levels of protection against demand variability.

If we carry no safety stock and have only the 130 rolls of cycle stock, that will be enough to satisfy all demand for product A on half the cycles; half the time demand will be at 130 rolls or less, and half the time greater. So with no safety stock, we will be vulnerable to stock-outs on half the cycles. Statistics teaches us that if we carry extra stock equal to one σ, that will be enough to cover demand on 84% of all cycles, as shown in Figure B.2. Sigma is 28 rolls for product A, so if we carry 28 rolls of safety stock in addition to the 130 rolls of cycle stock, that should be sufficient to prevent stock-outs on 84% of the cycles, about 44 weeks. If we carry safety stock equal to 2σ, that should cover 98% of the cycles, as shown in Figure B.3

Thus the key to determining safety stock is deciding on the tolerance for stock-outs, and then using that to determine how many sigmas of variability you need to cover. For example, if you decide that you can tolerate stock-outs on no more than 2% of the cycles, that sets the cycle service level goal at 98%, and we saw in Figure B.3 that that requires 2σ of safety stock, or 56 rolls. The percentage of cycles you hope not to have stock-outs is called cycle service level, and the

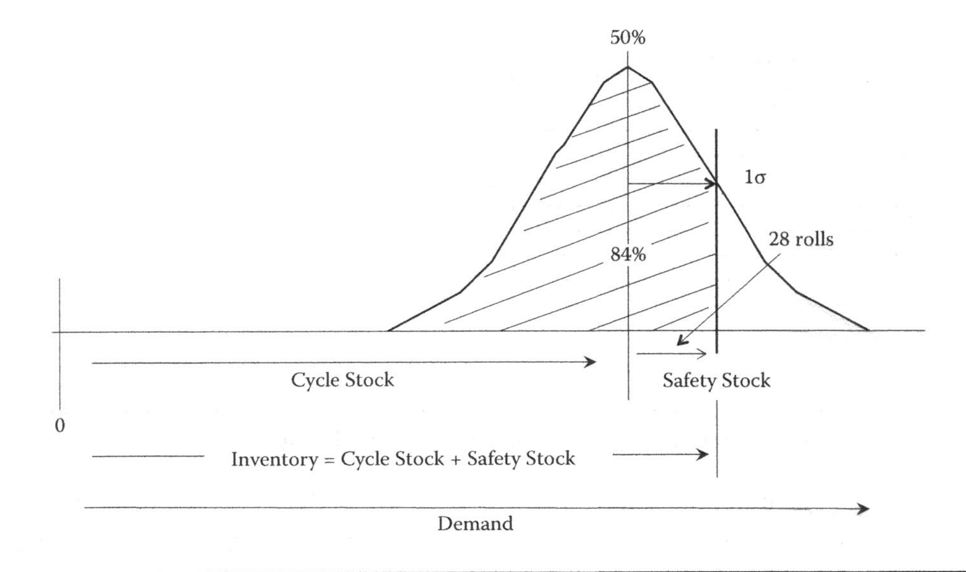

Figure B.2 Safety stock equal to one standard deviation covers 84% of the cycles.

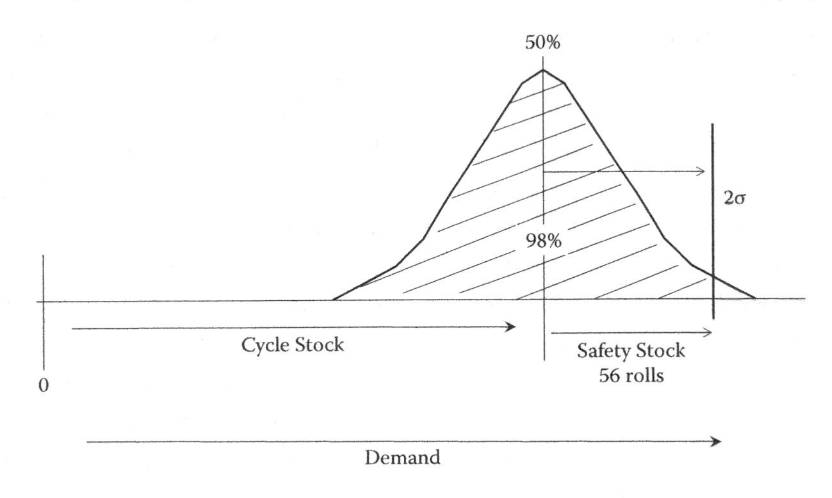

Figure B.3 Safety stock equal to two standard deviations covers 98% of the cycles.

number of sigmas required to achieve that is called the service level factor or the *Z* factor.

Thus the general equation for safety stock required to cover demand variability is

$$\text{Safety Stock} = Z \times \sigma_D$$

Figure B.4 shows the relationship between *Z* and service level. As can be seen, the relationship is highly nonlinear: higher service level values, i.e., lower potential for stock-out, require disproportionally higher safety stock levels. Statistically, 100% service level is impossible.

Typical service level goals are in the 90 to 98% range, but good inventory management practice suggests that rather than using a fixed *Z* value for all products, *Z* be set independently for groups of products based on strategic

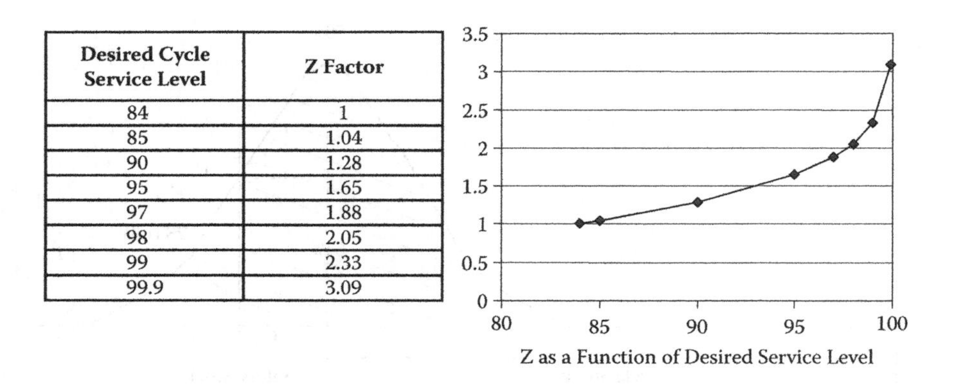

Desired Cycle Service Level	Z Factor
84	1
85	1.04
90	1.28
95	1.65
97	1.88
98	2.05
99	2.33
99.9	3.09

Figure B.4 Relationship between service factor and service level.

importance, profit margin, dollar volume, or some other criterion. Doing this will place more safety stock in those SKUs with greater value to the business, and less safety stock in the products believed to be less important to business success.

The above equation assumes that the standard deviation of demand is calculated from a data set where the demand periods are equal to the wheel time. If not, an adjustment must be made to the standard deviation value to statistically estimate what the standard deviation would be if calculated based on the periods equal to the wheel time. As an example, if the standard deviation of demand is calculated from weekly demand data, and the product wheel time is 2 weeks, the standard deviation of demand calculated from a data set covering 2-week periods would be the weekly standard deviation times the square root of the ratio of the time units, or $\sqrt{2}$. Bowersox and Closs, in *Logistical Management*, use the term *performance cycle* (PC) to denote the total lead time. If we let $T1$ represent the time increments from which the standard deviation was calculated (1 week in the above example), and PC to represent the wheel time plus any lead time needed to get to the inventory, then

$$\text{Safety Stock} = Z \times \sqrt{PC/T1} \times \sigma_D$$

With product wheels the performance cycle includes:

- The product wheel time.
- The time to get to the downstream inventory replenished by the wheel. There may be additional process steps between the wheel step and the next inventory storage location.
- Any review period included in the wheel scheduling process.
- If we are carrying inventory in a finished product warehouse, and customers allow a delivery lead time greater than the time needed to deliver to the customer, then the remaining customer lead time can be subtracted from the performance cycle.

The performance cycle can be considered the time at risk, i.e., the time between making a determination on how much to produce and making the next determination and having it realized.

Ideally, the exact amount to be produced on any spoke is finalized at the start of that spoke, when the most up-to-date inventory status is available. If the production quantity for all products is set at the start of the wheel cycle, the performance cycle for any spoke must include the time lapse between the start of the cycle and the start of the spoke, often called the review period. This review period adds to the time of demand uncertainty, so we will need more safety stock to cover the increased uncertainty. For example, a product made on day 3 of a 7-day wheel has a review period of 2 days, which must be added to the performance cycle. If we are about to start cycle 24, and we make a determination of how much of that product we will make on cycle 24, we are vulnerable to increases in demand on the full cycle 24 and the first 2 days of cycle 25. So we have a period of 9 days over which demand can vary before we can produce to accommodate it, and the safety stock must cover those 9 days of uncertainty.

Again, if the amount to be produced on any spoke is determined at the start of that spoke rather than at the start of the whole wheel cycle, the review period is zero. So in the above example, we are at risk of stock-out on only the last 5 days of cycle 24 and the first 2 days of cycle 25, for a total of 7 days of uncertainty.

If cycle stock has been calculated from historical demand, then the variance used in the safety stock calculation should be based on past demand variation. If forecasts are used to set cycle stock, then the factor requiring protection is forecast error. Standard deviation of forecast error would replace standard deviation of past demand in the safety stock formula, which would become

$$\text{Safety Stock } = Z \times \sqrt{PC/_{T1}} \times \sigma_{Fcst\,Err}$$

If there is bias in the forecast, it must be removed for the safety stock calculation to be valid. (Dealing with forecast bias is beyond the scope of this handbook.)

Variability in Wheel Time

The above equation calculates the safety stock needed to mitigate variability in demand. If variability in wheel time is of concern, that is, sometimes a wheel cycle runs long due to machine failures, temporary line upsets, or other problems, the safety stock needed to cover that is

$$\text{Safety Stock} = Z \times \sigma_{WT} \times D_{avg}$$

where *WT* is wheel time.

The average demand term (D_{avg}) is in the equation to convert standard deviation of wheel time expressed in time units into production volume units (gallons, pounds, rolls, etc.).

Combined Variability

If both demand variability and wheel time variability are present, the safety stock required to protect against each can under some conditions be combined statistically, to give a lower total safety stock than the sum of the two individual calculations. If demand variability and wheel time variability are independent, that is, the factors causing demand variability are not the same factors influencing wheel time variability, and if both variabilities are reasonably normally distributed, the combined safety stock is Z times the square root of the sum of the squares of the individual variabilities:

$$\text{Safety Stock} = Z \times \sqrt{{PC}/{T1}\ \sigma_D^2 + \sigma_{WT}^2 D_{avg}^2}$$

The idea behind this is that there is a low probability that you will see the highest demand and the longest wheel delay on the same cycle, i.e., that you will be at the extremes of both variability curves at the same time.

If σ_D and σ_{WT} are not statistically independent of each other, i.e., they are both influenced by the same factors, then this equation can't be used, and the combined safety stock is the sum of the two individual calculations.

$$\text{Safety Stock} = (Z \times \sqrt{{PC}/{T1}} \times \sigma_D) + (Z \times \sigma_{WT} \times D_{avg})$$

Using Safety Stock

We have defined safety stock to be inventory required to protect against variation in demand, variation in supply lead time, or both. Safety stock is being carried because we intend to use it, and will use it frequently. Since demand will be higher than average about half the time, we should expect to consume some of the safety stock during about half of the cycles, as depicted in Figure B.5. Further, if safety stock is calculated based on a 95% cycle service level, stock-outs should be expected on approximately 5% of the cycles.

Example—Forming Machine 2 Product Wheel

As an example of the use of the above calculations, consider the inventory levels required in the storage being replenished by forming machine 2, between it and

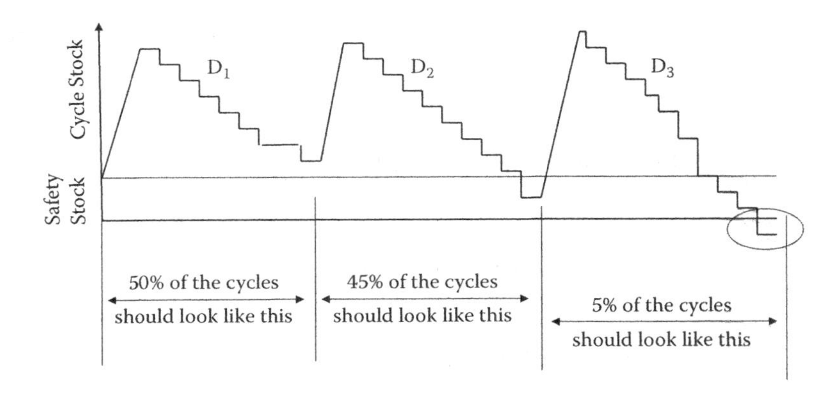

Figure B.5 With a 95% service level, stock-outs can happen.

bonder 2, shown earlier in Figure 5.3. As bonder 2 pulls a roll from this inventory, it creates a signal that is loaded onto the schedule for the product wheel on forming 2. When the appropriate spoke comes up on the forming product wheel, all material of that type gets formed, then flows into storage.

Looking specifically at the inventory requirements for product A, made on forming machine 2:

- Weekly demand = 130 rolls
- Standard deviation of weekly demand = 28 rolls
- Standard deviation of lead time = 0

Because forming is running a product wheel with a 7-day wheel time, and product A is being produced every cycle, the cycle stock for product A is 7 days' worth, 130 rolls.

The lead time is stable and very predictable; forming reliability is high enough that the wheel time never exceeds the 7 days. So with a standard deviation of lead time of zero, the safety stock requirement can be calculated from

$$\text{Safety Stock} = Z \times \sqrt{PC/T1} \times \sigma_D$$

If 95% is the desired cycle service level (the business can tolerate stock-outs of this product on no more than 5% of the replenishment cycles, slightly more than two per year), the Z value can be found in Figure B.4 to be 1.65. *PC*, the performance cycle, is the 7-day wheel time. $T1$, the time increments from which σ_D was calculated, is 7 days, so no correction to the standard deviation is necessary.

In this process, the amount to be produced on any spoke is determined at the start of that spoke, so the roll consumption inventory quantity is up to date, and the review period is zero. Thus

$$\text{Safety Stock} = 1.65 \times \sqrt{7/7} \times 28 \text{ rolls} = 46 \text{ rolls}$$

(If the service level goal had been 98%, Z would be 2.05 and safety stock would be 57 rolls.)

$$\text{Peak Inventory} = \text{Cycle Stock}\left(1 - \frac{D}{PR}\right) + \text{Safety Stock}$$

where D = 130 rolls/week and PR = 540 rolls/week.

$$\text{Peak Inventory} = 130 \text{ rolls}\left(1 - \frac{130}{540}\right) + 46 \text{ rolls} = 145 \text{ rolls}$$

$$\text{Average Inventory} = \frac{1}{2}(\text{Cycle Stock})\left(1 - \frac{D}{PR}\right) + \text{Safety Stock}$$

$$\text{Average Inventory} = \frac{1}{2}(130 \text{ rolls})\left(1 - \frac{130}{540}\right) + 46 \text{ rolls} = 96 \text{ rolls}$$

When the spoke for product A is being run on the product wheel, it should make enough to bring the A inventory back up to the 145-roll level by the end of its production. This will require production of 130 rolls plus any safety stock that was consumed, since 31 rolls will be consumed while it is being produced.

If there were variability in lead time, if forming equipment failures caused the 7-day wheel time to vary, more safety stock would be required to meet the inventory performance goals. If, for example, wheel time varied with a standard deviation of 1/2 day, or 0.07 week, the safety stock calculation would be as follows:

$$\text{Safety Stock} = Z \times \sqrt{\frac{PC}{T1}\ \sigma_D^2 + \sigma_{WT}^2 D_{avg}^2}$$

$$\text{Safety Stock} = 1.65 \times \sqrt{\left(\frac{7}{7}\right)28^2 + (0.07)^2 130^2}$$

$$\text{Safety Stock} = 1.65 \times \sqrt{784 + 83} = 49 \text{ rolls}$$

Two things can be seen from this result. The first is that in this example the demand variability has the dominant influence on safety stock requirements: its effect on safety stock is almost 10 times that of lead time variability. It is often the case that one factor or the other will dominate the calculation; it is important to recognize that, so that improvement efforts can be focused on the most appropriate things. In this case, if we decide to reduce the need for safety stock, it is far more productive to work on demand variability than on lead time variability.

The second observation is that the influence of lead time variability is so small that safety stock requirements increase by only 3 rolls, to 49 rolls, compared to what it was without considering lead time variation.

To summarize, given the parameters above, if the inventory for A includes 46 rolls of safety stock, or 49 rolls in the case of both variabilities, stock-outs of product A should be expected to occur in no more than 5% of the cycles, or during two to three of the weekly product wheel cycles each year. With the same service goals, stock-outs of each of the other products should also be expected on 5% of the cycles.

If that number of stock-outs is unacceptable, then higher service level goals should be set, resulting in higher *Z* values and more safety stock required. One key advantage of these calculations over arbitrary rules of thumb is that the trade-offs can be quantified, so the business can make informed decisions on how it wants to balance inventory cost against risk of stock-outs.

Appendix C: Total Productive Maintenance

The Need for Equipment Reliability and Operational Continuity

Reliable equipment, operating stably and predictably, is a key requirement for effective product wheel operation. If equipment suffers from reliability issues, you may not be able to depend on completing the wheel cycle within the planned wheel time.

Well-maintained equipment contributes to product wheel performance in three ways:

1. Lower downtime means more time available for changeovers, so wheel time can be shorter.
2. Lower downtime means more time can be available for process improvement time activities—equipment modifications and upgrades, new product development and qualification runs, operator training, etc.
3. Higher uptime means higher effective capacity and higher throughput, important if the process is capacity constrained or nearly so.

Total productive maintenance (TPM) is a very effective way to improve equipment reliability and ensure equipment continuity, and thus provides a very critical foundation for product wheels.

TPM

Total productive maintenance is a philosophy, a set of principles, and specific practices aimed at improving manufacturing performance by improving the way that equipment is maintained. It was developed in Japan in the 1960s and 1970s, based on preventive maintenance (PM) and productive maintenance practices developed in the United States. But where PM is focused on the maintenance shop and on mechanics, TPM is team based and involves all parts and all levels of the organization, including supervisors, plant managers, and perhaps most importantly, operators. It drives toward autonomous maintenance, where the majority of maintenance is done by those closest to the equipment, the operators. In this operating model, the maintenance group can now focus on equipment modifications and enhancements to improve reliability.

The goal of TPM could be described as the development of robust, stable value streams by maximizing overall equipment effectiveness (OEE).

Some key elements of TPM are:

1. Preventive maintenance—Time-based maintenance, maintenance done on a schedule designed to prevent breakdowns before they can occur. Very effective when failures occur at a relatively predictable component life.
2. Predictive maintenance—Condition-based maintenance, using instruments and sensors, such as heat and vibration monitors, to try to anticipate when equipment is about to break down so that it can be fixed before failing.
3. Breakdown maintenance—Repairing the equipment after a breakdown occurs. Often the only viable alternative if failures occur randomly and unpredictably.
4. Corrective maintenance—Ongoing modifications to the equipment to reduce the frequency of breakdowns and make them easier to repair.
5. Maintenance prevention—Design equipment that rarely breaks down and is easy to repair when it does fail.
6. Autonomous maintenance—Team-based maintenance done primarily by plant floor operators. This is perhaps the most powerful element of TPM. Teaching operators maintenance skills so that they can fix routine disruptions and failures leads to faster response and more timely corrections and repairs. Operators are in the best position to recognize impending failures: they are well tuned in to how the equipment should sound, smell, and feel.

The most fundamental element of TPM is the culture change required, moving from a mindset that the maintenance group owns accountability for equipment performance to one where everyone in the plant has that accountability.

TPM Metric—Overall Equipment Effectiveness

One of the most widespread measures used to gage the effectiveness of a TPM effort is OEE. One reason for its popularity is that it captures in a single metric all of the factors that detract from optimum equipment performance.

OEE is the product of three factors (Figure C.1):

- Availability
- Performance
- Quality

Availability captures all downtime losses, including breakdown maintenance, preventative maintenance, and time spent in changeovers. Availability is calculated as actual operating time divided by planned production time.

Note that setup time or changeover time should not include the time getting properties back within specification after the changeover, since that loss is captured in the quality factor in OEE.

$$\text{Availability} = \frac{\text{Actual Operating Time}}{\text{Planned Operating Time}}$$

Performance captures the loss in productivity if equipment must be run at less than the design throughput rate because of some equipment defect. For example, chemical batches can take longer to heat up or react if residue has built up on vessel walls, thus impeding heat transfer. Rotating machinery, paper winding equipment, or plastic film processing equipment may have to be run at slower speeds if bearings are worn. Performance is calculated as actual throughput divided by rated throughput.

A caution on performance calculation: in the process industries, there are often rate limitations due to the requirements of the material being processed, not due to any equipment defect. For example, when heat-treating sheet goods to set properties in, some products may require that the heat treater be run slower to allow for more time at temperature than required for other products. Likewise, some batches of paint resin may require more "cook time" to reach the desired viscosity level than other resins with different target viscosities run on that equipment. Since these rate limitations are due to product requirements rather than equipment performance, they are truly value adding, so the performance metric should not be penalized.

$$\text{Performance} = \frac{\text{Actual Throughput}}{\text{Rated Throughput}}$$

(Actual throughput and total throughput should both be weighted averages for equipment that must run at different rates for different products.)

Quality captures the loss in equipment productivity when out-of-specification product is being made, including scrap material, material that must be reworked to be acceptable, and yield losses during start-up or when coming back from a product changeover.

$$\text{Quality} = \frac{\text{Quantity of First Grade Material}}{\text{Total Quantity Produced}}$$

OEE is then calculated as:

$$\text{OEE} = \text{Availability} \times \text{Performance} \times \text{Quality}$$

Uptime is another metric often used to gage the effectiveness of TPM initiatives. It measures the same losses that OEE does, but performs the calculations in a different way. Even so, it gives the same result as the OEE calculation.

Forming 2 OEE

The OEE on forming 2 prior to product wheel implementation was 79%. This was based on an Availability of 87%, Performance of 100%, and Quality of 91%.

- 9-day cycle
- 14 products on each cycle
- Hours per cycle = 9 × 24 = 216 hours
- Changeover time = 30 minutes (without restart time loss of one roll—that is accounted for in quality)
- Reliability = 90%
- Yield = 91%
- Availability = (216 – 21.6 – (14 × 0.5))/216 = 87%
- Performance = 100% (there are no rate reductions due to equipment problems)
- Quality = 91%
- OEE = 87% × 100% × 91% = 79%

The OEE improved slightly with the start-up of the wheel. The number of changeovers went down, from 14 changeovers per 9-day cycle to an average of 7 changeovers on each 7-day wheel cycle. Because slightly less time was taken up with changeovers, the Availability went up by 1%, and because fewer restart rolls were lost, Quality increased by ½%.

- 7-day cycle
- An average of 7 products on each cycle
- Hours per cycle = 7 × 24 = 168 hours
- Changeover time = 30 minutes (without restart time loss of one roll—that is accounted for in quality)
- Reliability = 90%
- Yield = 91.5%
- Availability = (168 – 16.8 – (7 × 0.5))/168 = 88%
- Performance = 100% (there are no rate reductions due to equipment problems)
- Quality = 91.5%
- OEE = 88% × 100% × 91.5% = 80.5%

Thus the application of a product wheel to forming 2 has increased OEE by 1.5%, largely by decreasing the number of changeovers per unit time.

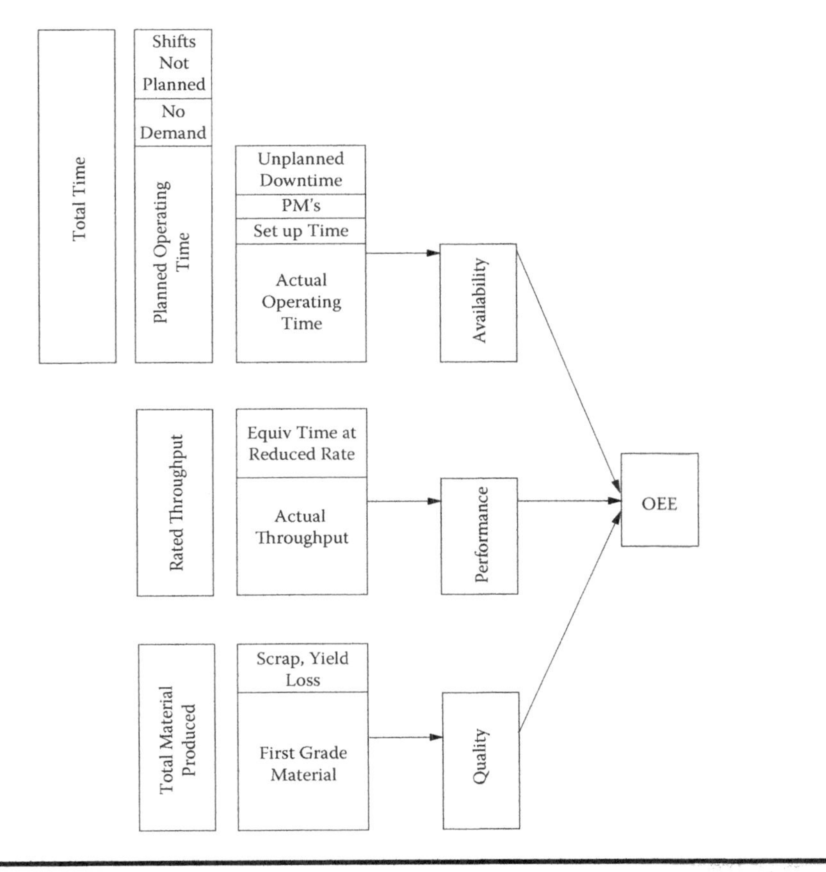

Figure C.1 The components of OEE.

Appendix D: The SMED Changeover Improvement Process

The time consumed and material losses on product changeovers are among the primary determinants of overall product wheel time and product sequencing, so any wheel design and implementation program must place a heavy emphasis on changeover improvement. Difficult and costly changeovers drive long production campaigns and high inventories; if changeover time can be reduced, wheel times can be reduced. And if any material losses experienced in changeovers, either in material flushed at the beginning of a changeover or in material lost on restart while getting back to product specifications, can be reduced, it will drive the economic order quantity (EOQ) calculation toward shorter cycles.

SMED (single-minute exchange of dies), conceived by Shigeo Shingo while working at Toyota in the 1950s and then perfected by Toyota over the next 30 years, has become a widely used best practice for changeover simplification and reduction.

SMED Origins

As the Toyota Production System was beginning to evolve in the early 1950s, Toyota recognized that it was critical that product changes be accomplished as quickly as possible so that short campaigns would be feasible.

One of the most time-consuming changeovers they faced was the replacement of the dies on the large hydraulic presses used to stamp out auto body parts. Shigeo Shingo, an industrial engineer consulting with Toyota, developed a methodology for examining all setup operations and modifying the setup process to reduce the overall time. Using Shingo's techniques, Toyota was able to shorten the die changes from 3 hours in the 1950s to 15 minutes by 1962, and to an average of 3 minutes by 1971 (Ohno, 1988). In recognition of this tremendous accomplishment, Shingo's methods and techniques have become the standard for changeover reduction and have come to be known by the acronym SMED.

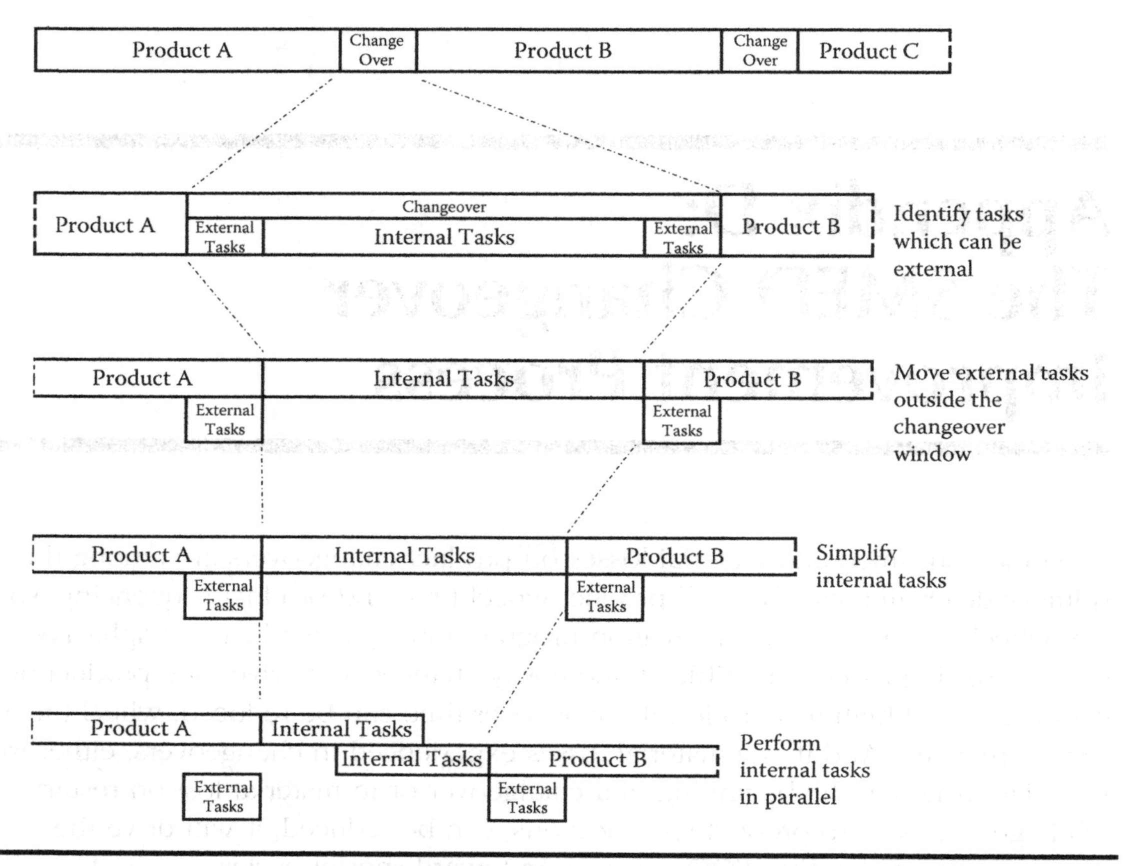

Figure D.1 Four steps in the SMED process.

SMED Concepts

Figure D.1 shows the four steps in the SMED process.

1. Determine if any of the tasks done during the changeover can be done before the equipment is turned off and production stopped or after the equipment is turned back on and making good product. These tasks can consume a lot of time, so moving them outside of the time window when the machine is not producing can shorten changeover time dramatically. Any tasks that must be done during the changeover are called internal tasks, while those that could be done before or after are called external tasks. So step 1 is to identify any external tasks.
2. Move external tasks outside of the changeover time.
3. Simplify the remaining internal tasks.
4. Where feasible, perform internal tasks in parallel. If several operators can perform tasks concurrently, the time can be reduced without increasing the total labor content of the setup.

After the changeover process has been revised and tested, it is critical that it be documented, standardized, and audited on an ongoing basis so that the improvements can be sustained.

It is also critical that if a wheel is already in place, wheel time then be reexamined, and shortened to take advantage of the changeover time reduction.

The calculations done in wheel design provide a basis for demonstrating the economic benefit of any changeover improvements, which will be necessary to get funding if the improvements require any capital investment.

Product Changeovers in the Process Industries

SMED is particularly valuable on process industry product wheels, because the changeovers tend to be much more complex and involve more costly material losses.

In process operations, a lot of time may be spent in cleaning out the raw material feed systems and the processing equipment to prevent cross-contamination. For example, the tinting tanks used in paint manufacturing require thorough cleaning during color changes. In many food processing plants, equipment is shared among several product varieties, which may or may not contain allergens, such as peanuts. This can pose very stringent requirements for cleaning between products, and then extensive testing to ensure that the equipment is free of contaminants. The cleanouts and testing required in pharmaceutical manufacturing are even more thorough. The tasks performed during these cleanups are very well suited to SMED analysis.

In extrusion, sheet good, and batch chemical processes, much of the time lost is the time required to bring the line to the appropriate temperature, pressure, speed, thickness, etc., after all the mechanical tasks have been performed. Therefore the SMED process is very helpful in attacking these so that the total time for the changeover, including the time for process conditions to stabilize, is reduced.

Summary

If time and material losses during changeovers can be reduced, better product wheels will result. Ideally, SMED should be done before product wheel design begins, but if wheels are already in place, SMED improvements will be just as helpful in improving wheel performance.

SMED is not a one-time event. Experience in many applications has shown that repeating the process periodically will surface opportunities that weren't found previously. Toyota's experience is a dramatic example: it took the company 20 years of relentlessly repeating the SMED process, but as a result, it went from 3 hours to 3 minutes!

Appendix E: Bottleneck Identification, Improvement, and Management

A bottleneck is any piece of equipment or step in a process that has effective capacity less than or equal to takt. In other words, it is a step that cannot meet, or can barely meet, the demand placed on it. If there is a bottleneck anywhere in the line on which you are using a product wheel, it will limit capacity and make it difficult or impossible to follow the wheel schedule.

Finding and opening up any bottlenecks is therefore essential to successful wheel operation. If there are bottlenecks upstream of the product wheel step, flow of necessary material to the wheel step will be limited and the wheel will not perform well. If there is a bottleneck downstream of the wheel step, with no inventory in between, the wheel will have to stop periodically as the bottleneck limits flow.

If there are steps that are not bottlenecks, but are nearly so, opening them up creates additional usable capacity and enables smoother wheel operation.

In discrete parts manufacturing and assembly plants, bottlenecks are often the result of poor labor staffing, so bottleneck management is a matter of better manpower scheduling practices. Bottlenecks in process plants tend to be much more related to equipment capacity and performance than to labor and staffing, so bottleneck management must focus on asset performance.

Root Causes of Bottlenecks

The most common reasons for bottlenecks in process plants will include:

1. Inherent equipment capacity limitations
2. Long changeovers reducing usable capacity
3. Mechanical reliability problems
4. Yield losses
5. Inappropriate scheduling

In *Synchronous Manufacturing*, Umble and Srikanth (1990) use the term *capacity constraint resource* to distinguish a step in the process that can't meet demand not because of any inherent equipment limitation, but due to the way it is being scheduled. If all of the steps in a process are not well coordinated and synchronized, that by itself can limit flow and create an apparent bottleneck.

The data boxes on the value stream map (VSM) should help identify any bottlenecks in your process, and give an indication on why each is a bottleneck, another good reason why we need to start with an accurate, up-to-date VSM. For example, looking at the VSM for Cells 1 and 2 of our sheet goods process (Figure 5.3), we see that we have no real bottlenecks. The forming and bonding machines all have utilizations in the 68–85% range, indicating that they have reasonable capacity beyond that required to make takt. The equipment that is closest to being a bottleneck is slitter 1, with a utilization of 87%. Attempts to open that near bottleneck up would include a SMED analysis (see Appendix D) of the methods used to reposition the cutter knives on product changeovers.

Bottleneck Management—Theory of Constraints

After any bottleneck has been identified and the root cause diagnosed, it's time to open it up, i.e., to make whatever changes that can be identified to resolve the bottleneck. At the same time, it is important to make sure that throughput at the bottleneck is not suffering unnecessarily from problems upstream or downstream of the bottleneck.

It may require inventory to accomplish this, but adding inventory is often the most reasonable compromise when compared to the alternative of not being able to make takt.

The strategies for managing and optimizing bottlenecks described by Eli Goldratt in his two landmark works, *The Goal* (Goldratt and Cox, 1984) and *Theory of Constraints* (Goldratt, 1990), have become the standard for dealing with them. The process laid out by Goldratt can be summarized as follows:

1. Identify the bottleneck.
2. Exploit the bottleneck: Make sure that the bottleneck is running at maximum capacity, and not wasting time on noncritical tasks.
3. Subordinate everything else to the bottleneck: All upstream and downstream processes should operate in a way that maximizes bottleneck throughput.
4. Elevate the bottleneck capacity: Try to increase the capacity of the bottleneck.
5. Once the bottleneck is broken, find the next bottleneck and repeat.

Step 3 is perhaps the most important and most often ignored step. It suggests placing buffer inventories before and after the bottleneck step, so that the bottleneck will not be further constrained by upstream or downstream disruptions.

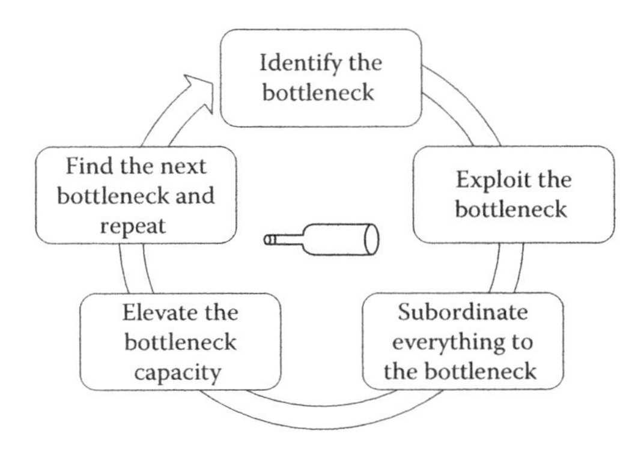

Figure E.1 Bottleneck management process.

The buffer before the bottleneck step will provide material to the bottleneck even when upstream processes are down. The buffer after the bottleneck step provides a place to hold the bottleneck output when downstream process steps are down. These actions won't open the bottleneck up at all, but they will ensure that flow through the bottleneck is not further limited.

Summary

A fundamental premise of product wheel design is that the line has sufficient capacity to meet demand. If that condition is not met, you will find the wheel frequently falling behind schedule, and so all effort on implementing wheels will have been wasted. For that reason it is important to understand if there is a bottleneck or near bottleneck in your line, which can be seen from the utilization figures on the VSM. Once a bottleneck is found, it should be opened up as much as possible, using SMED, TPM, and other practices described throughout this handbook. You should also make sure that uncoordinated scheduling practices aren't creating a capacity constraint. And then the bottleneck should be protected against further flow limitations using Theory of Constraints thinking.

Appendix F: Group Technology and Cellular Flow

Of all of the improvement tools in the Lean toolbox, cellular manufacturing is perhaps one of the most powerful. It enables smaller lot production, more visible flow, quality improvements, reduced work in process (WIP), shorter lead times, and simplifies the implementation of product wheels and pull replenishment systems.

With process layouts where cellular manufacturing is applicable, having it in place can greatly simplify product wheel design. The cellular concept separates material flow paths and product groupings into much more manageable subsets, meaning that each wheel will have fewer products and fewer variables to manage.

Typical Process Plant Equipment Configurations

Cellular manufacturing only applies where you have similar equipment in parallel at one or more points in the process. It is particularly beneficial where there is parallel equipment at several steps, like the configuration shown in Figure F.1, a pattern typical of many process industry plants. There is a small number of key processing steps, in this case four, and there are a few (three or more) machines, tanks, or reaction vessels in parallel at each step. The parallel machines are quite similar, and often a specific material can be processed by any one of them. Occasionally, the machines or vessels have some unique capabilities such that some materials must go to a specific machine or vessel.

Figure 4.1 showed this configuration for our sheet goods process. There are four very similar roll forming machines, four similar roll bonders, three slitting machines, and three cutters or choppers. A roll produced on any of the four forming machines may be processed by any of the bonders, slitters, and choppers.

Process plants usually require this array of equipment to handle the high volume of material to be produced. Practical equipment size limitations prohibit the design of a single machine or vessel large enough to process the full throughput required. Product mix considerations, i.e., the high degree of product variety, would encourage the use of many small machines to give more flexibility, but economies of scale have deterred plant designers from going in that direction. As

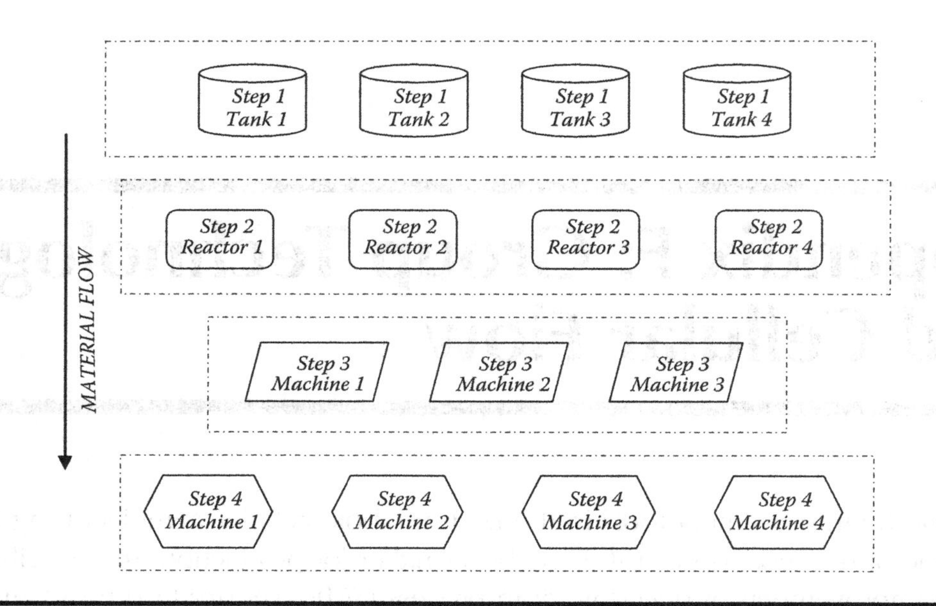

Figure F.1 Typical process industry equipment footprint—Functional configuration.

capital cost optimization has traditionally overridden good Lean thinking in process plant design, the result is a few large vessels or machines at each process step.

This equipment configuration is highly valued for the flexibility it offers. If a batch of material is leaving step 1, and one of the step 2 machines is down for maintenance, there may be others available to process the material. The result is that flow paths are often as shown in Figure F.2. All of the flexibility inherent in the system is exploited, but generally with more negative than positive consequences. There is frequently a belief that utilizing this flexibility maximizes asset utilization, although the opposite is usually true.

This mode of operation, taking advantage of the inherent flexibility of this configuration, brings a number of problems. Material tends not to flow directly from one step to the next, but to be put into some type of storage. In the sheet goods process, formed rolls typically do not flow directly to a bonder: instead, they are taken to an automatic roll storage system, to be retrieved later for transport to a bonder. Thus large WIP storages are created. Flow becomes unsynchronized, is difficult to visualize, and is even more difficult to manage. Because each piece of equipment can process any of the product types, each product wheel has the entire product lineup to consider.

Quality suffers for two reasons. There is a significant time lapse between each process step, so any defects or out-of-spec material may not be discovered for some time, making all of the intermediate material suspect. Even with this simple-looking arrangement, there are 192 (4 × 4 × 3 × 4) possible flow path combinations. Since no two machines or vessels will produce exactly the same product, that provides 192 different ways that process variabilities can add up. A statistical process control (SPC) specialist would tell you that you don't have a process, you have 192 different processes. With so many variables, root cause analysis of product defects becomes very difficult.

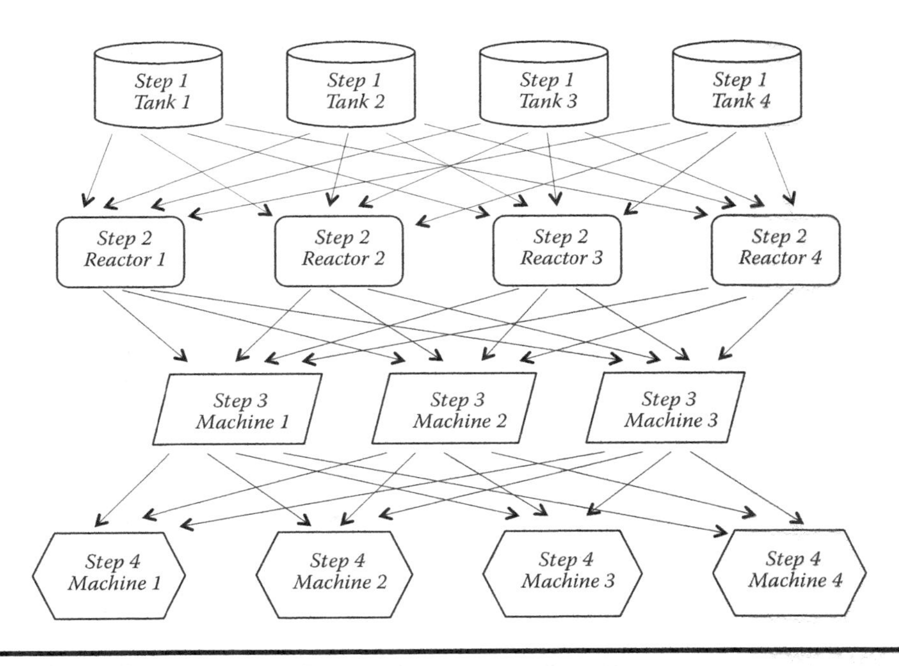

Figure F.2 Typical process industry flow patterns.

Because there are thought to be alternate paths available whenever a piece of equipment fails, there is far less urgency to maintain the equipment appropriately. Thus with time, equipment performance as measured by overall equipment effectiveness (OEE) or Uptime deteriorates.

Cellular Manufacturing Applied to Process Lines

These problems can all be mitigated by applying cellular manufacturing concepts.

In discrete parts manufacturing processes, cellular manufacturing has traditionally required relocating the equipment into U-shaped or L-shaped patterns, to provide shorter paths for operators to travel. It has therefore been thought that cells weren't applicable to most process operations, because the equipment is most often very large and expensive to relocate. But the advantages of cellular flow can be achieved without moving anything, by creating virtual flow paths, so that flow is managed in a cellular fashion. The key is to think in terms of **flow** rather than **function**.

The basic concept is straightforward: start by grouping all process materials or products into families requiring similar process conditions. Then identify the process equipment required by each family, but instead of creating a work cell by rearranging the equipment, create virtual work cells by defining the acceptable flow patterns. Figure F.3 shows what this would look like for the process diagrammed in Figures F.1 and F.2. Again, no equipment would have to be moved; the new, more limited flow patterns would simply have to be defined and followed.

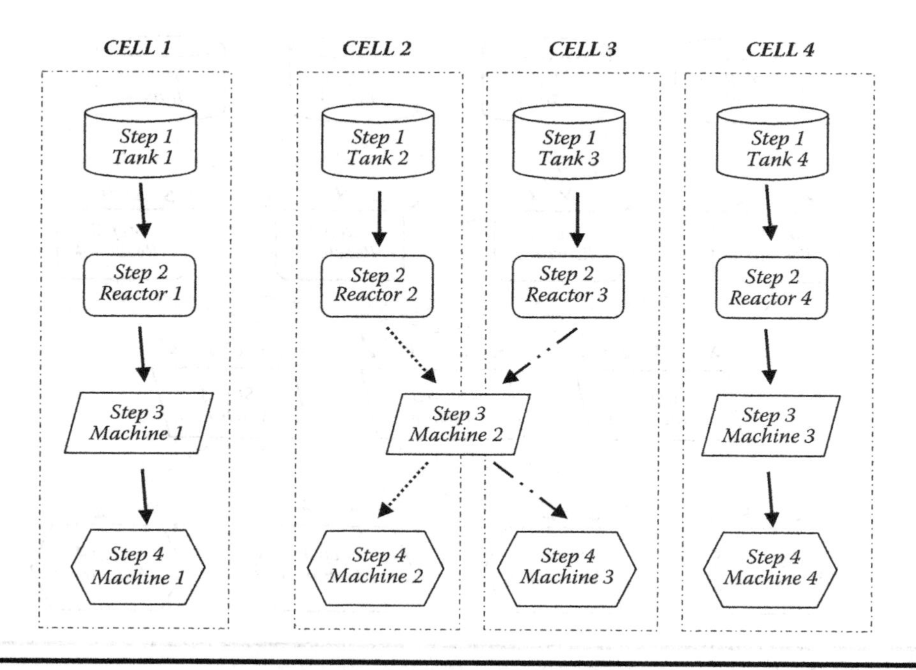

Figure F.3 Grouping into virtual work cells.

The advantages of this virtual work cell concept are:

- Flow becomes far easier to understand, visualize, and manage.
- Flow tends to be more continuous, with less material being transported to storage, so WIP and material handling are reduced.
- Quality improves because feedback is much more immediate.
- As depicted in Figure F.3, we now have only four possible flow paths instead of the 192 we had before, so product variability is dramatically reduced.
- As explained below, each cell is generally processing a subset of the full product mix with similar requirements, so changeovers become far easier and usable capacity increases.

It must be recognized that the numbers don't always work out perfectly, but that reasonable compromises can usually be found that will give most of the benefit. In the case shown, since there are only three machines at step 3, one must be shared between cells 2 and 3. If that machine didn't have enough capacity to process the total throughput of the two cells, additional compromises would have to be made, perhaps as shown in Figure F.4, with six possible flow path combinations. But comparing Figure F.4 to Figure F.2, you can see that even in the worst case, flow paths will be far reduced from the noncellular flow.

Figure 5.1 showed a virtual cell layout for our sheet goods process. Because there are only three slitters, slitter 1 would be shared between cells 1 and 2. Since the choppers are designed to cut from different incoming slit widths, all

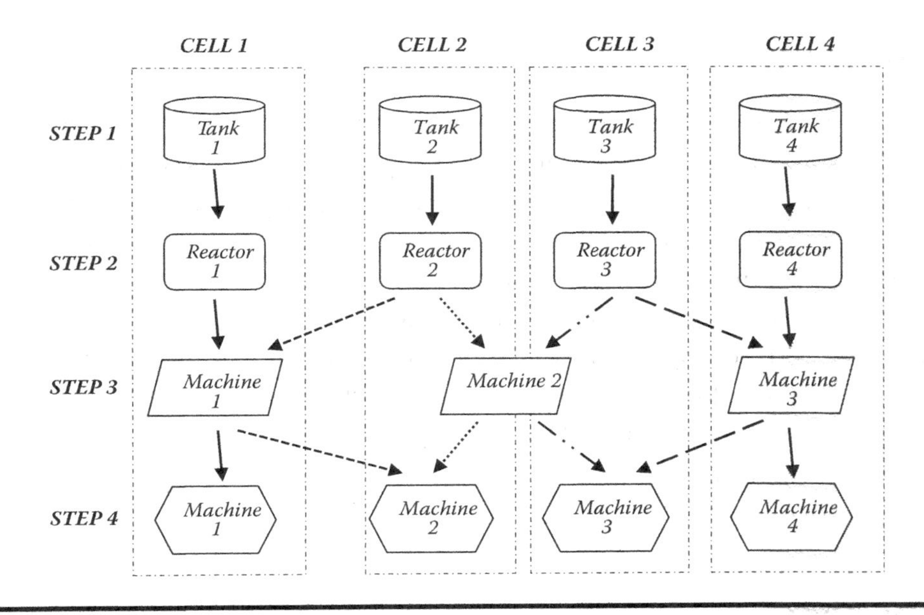

Figure F.4 Alternate virtual work cell grouping.

three are required by each cell, so the choppers are not included in the virtual cells. Even with that, the flow is restricted to 12 possible combinations, where prior to the formation of virtual cells, 144 possible paths (4 × 4 × 3 × 3) existed. Thus the key advantages of cellular manufacturing apply.

Although the new flow paths are the more obvious aspect of cellular manufacturing, the more significant benefit is that the product portfolio can be divided up into families, so that each piece of equipment must process far fewer product types. Having fewer products on each wheel dramatically simplifies wheel design and operation.

An even more profound benefit is that the products assigned to a specific piece of equipment are usually grouped by similar processing characteristics, so that the changeovers on a piece of equipment are far faster and less expensive. For example, if the products assigned to a specific bonder are all heat-set within a narrow range of temperatures, the time required to change temperature on product changeovers is far less.

The part of the cellular manufacturing process where products are grouped into families with like characteristics and processing requirements is often called group technology. (The *APICS Dictionary* defines group technology as "an engineering and manufacturing philosophy that identifies the physical similarity of parts (common routing) and establishes their effective production. It … facilitates a cellular layout.") Even if the equipment can't be arranged into virtual cells, or if there is parallel equipment at only one step, group technology still offers significant advantage for that parallel equipment, by dividing the product lineup so that each of those parallel pieces of equipment has a smaller product mix to process.

Summary

Where the equipment configuration lends itself to cellular manufacturing, it should be designed and implemented before product wheels are considered. It should be apparent that wheels are far easier to put in place with the small number of flow paths that a cellular arrangement requires, and more importantly, with a reduced number of products to be processed on each piece of equipment. And where the configuration does not lend itself to cells, the concept of group technology should be applied wherever there is similar equipment in parallel. Not only does this reduce the number of products on each wheel, but a grouping can usually be found that reduces changeover time and cost, leading to much more effective wheels.

Index